The Biological Activity of Phytochemicals

Volume 41

David R. Gang

Editor

The Biological Activity of Phytochemicals

 Springer

Editor
David R. Gang
Washington State University
Pullman, WA 99164-6340, USA
gangd@wsu.edu

ISBN 978-1-4419-6961-3   e-ISBN 978-1-4419-7299-6
DOI 10.1007/978-1-4419-7299-6
Springer New York Dordrecht Heidelberg London

Library of Congress Control Number: 2010937099

Printed on acid-free paper

Springer is part of Springer Science+Business Media (www.springer.com)

Preface

The publication of this volume marks the reintroduction of the *Recent Advances in Phytochemistry* (*RAP*) series, an annual journal supported by the Phytochemical Society of North America. In the past, *RAP* was a proceedings book representing the focus of the PSNA annual meeting, and production of the annual RAP volume was largely the responsibility of an editor in chief with the assistance of the local organizing committee of the annual Society meeting. Recent years have brought significant changes to biological science in general and phytochemically related fields in particular. As a result, the annual meetings of the PSNA are no longer centered on a single theme, but rather represent the most interesting advances and newest technologies developed by the membership of the Society in the broadest sense. The meetings are now organized to be of interest to all members of the Society, whether they study the biosynthesis of natural products or regulation of metabolism, the ecology of specialized metabolites or the evolution of their pathways, or the effects of natural products or plants on human health, or whether they are involved with genomics, proteomics, metabolomics, natural product structural determination or new technology development, medicinal chemistry or metabolic engineering, or any of the myriad of fields that are now closely associated with what may be called "traditional phytochemistry" and plant biochemistry. The advent of post-genomics-based ways of thinking, of systems biology, of synthetic biology, of comparative genomics/proteomics/transcriptomics/metabolomics, and especially of the introduction and establishment of a mentality that leads to support of large collaborative projects, has opened up many new doors to scientists interested and versed in the (bio)chemistry of plants. In response to these changes, an Editorial Board was established to oversee the reintroduction and future production of the *RAP* series.

With the reintroduction of the *RAP* series, the Editorial Board decided to include two main types of articles in the journal format: Perspectives and Communications. Perspectives in *RAP* are expected to synthesize results from the primary literature and perhaps from new/novel results and place these in perspective relative to the broader field. These articles may be similar to review articles, but also are intended to present important ideas and hypotheses, and may present proposals for interesting directions in the field. It is the hope of the Editorial Board that these articles will be of great value to a large audience. Communications are intended to represent new advances in the field that will be of interest to a large audience. Articles of both types are typically solicited from the Society membership based on the content of the annual meeting talks, but in keeping with the title "Recent Advances in Phytochemistry," the editorial board reserves the right to solicit additional Perspectives and/or Communications from non-attendees as well (e.g., where an editorial board member has knowledge of an interesting recent advancement that would be of general interest to the society membership).

All submissions to *RAP* go through a rigorous peer review process, overseen by the Editorial Board, which includes external review. Starting this year, all *RAP papers* will be published electronically on SpringerLink.com with *Springer* journals as well as in book format.

This marks a significant change from past volumes of *RAP*, and it is the hope of the Editorial Board that this will lead to broader dissemination of the contents of and greater interest in *RAP*.

This 41st volume of RAP includes a total of 12 articles, all based on talks presented at the 49th annual meeting of the PSNA. These eight Perspectives and four Communications give a very good picture of the state of plant (bio)chemistry research in North America, which is also indicative of the state of the field worldwide. Each of these articles describes the integration of several different approaches to ask and then answer interesting questions regarding the function of interesting plant metabolites, either in the plant itself or in interactions with the environment (natural setting or human health application).

The first three Perspectives have a strong ecological focus. Tholl and Lee describe how volatiles such as homoterpenes are involved in ecological interactions between plants such as *Arabidopsis thaliana*, their herbivores and *their* parasites, such as by attracting parasitic wasps and mites in tritrophic interactions. Pedras outlines how natural products mediate molecular interactions between crucifers (Brassicaceae) with their pathogenic fungi, and how specific compounds are utilized by specific fungi in these interactions. Bernards et al. describe how complex metabolites such as the ginsenosides, a class of triterpenoid saponins, act to deter attack by non-pathogens but are potentially utilized by some pathogens as host recognition factors.

The next three Perspectives focus more on the biosynthesis of specific classes of natural products, although the roles of these compounds in their respective plants are also delineated. Umezawa et al. explain the importance of control of enantioselectivity in the biosynthesis of large numbers of specialized metabolites derived from the lignan, neolignan, and norlignan pathways. Guzman et al. describe recent advances in our understanding of color (e.g., carotenoid, flavonoid, chlorophyll), heat (capsaicinoid), and flavor (terpenoid, aldehyde) production in peppers. Owens and McIntosh outline what is known about flavonoid production in *Citrus* species, especially regarding the role that glycosylation plays in production of these compounds in these species, and make a strong case that future work on glycosyl transferases will be an important springboard for any efforts to alter flavonoid metabolism in these plants.

The Perspectives by Maschek et al. and by Stevens and Reed are more application oriented. Maschek et al. describe efforts to identify antiviral compounds from Antarctic red algae, and outline progress and challenges in following up on strong leads when very active compounds appear to be present at low levels. Stevens and Reed outline development and optimization of a fermentation procedure that resulted in efficient production of glucolimnanthin-derived isothiocyanate and nitrile natural herbicides from Meadowfoam (*Limnanthes alba*) seed meal, and they describe analytical methods used to characterize these compounds from complex sample extracts.

Four Communications are also included in this volume. McIntyre et al. describe how variation in ginsenoside content in Ontario-grown North American ginseng (*Panax quinquefolius*) is correlated with anti-glycation and antioxidant activities and propose that identifying variation in ginsenoside profiles is important for quality control and landrace development of ginseng. Kovinich et al. provide a perspective on anthocyanin biosynthesis in black soybean and potential strategies for engineering seed colors in light of substantial equivalence. Sharma et al. outline an efficient method to extract and fractionate proanthocyanidins from grape seed that can be achieved on a scale suitable for generation of sufficent material for bioactivity assays and animal feeding studies. Finally, Luu et al. describe how ginsenosides in ginseng extracts inhibit CYP3A4 but not CYP2C9 in vitro, suggesting that CYP3A4 inhibition by ginsenosides warrants further study in a clinical setting.

We hope that you will find these Perspectives and Communications to be interesting, informative, and timely. It is our goal that *RAP* will act not only as the voice of the PSNA, but that it will serve as an authoritative, up-to-date resource that helps to set the gold standard for thought and research in fields related to plant biochemistry.

We welcome suggestions for future articles and comments on the new format.

The RAP Editorial Board

<table>
<tr><td>Pullman, WA</td><td>David R. Gang</td></tr>
<tr><td>London, ON</td><td>Mark A. Bernards</td></tr>
<tr><td>Pullman, WA</td><td>Laurence B. Davin</td></tr>
<tr><td>Vancouver, BC</td><td>Reinhard Jetter</td></tr>
<tr><td>Peoria, IL</td><td>Susan McCormick</td></tr>
<tr><td>Tampa, FL</td><td>John T. Romeo</td></tr>
<tr><td>Towson, MD</td><td>James A. Saunders</td></tr>
</table>

Contents

Contributors

Charles D. Amsler Department of Biology, University of Alabama at Birmingham, Birmingham, AL 35294, USA, amsler@uab.edu

John T. Arnason Department of Biology, University of Ottawa, Ottawa, ON, Canada, john.arnason@uottawa.ca

Bill J. Baker Department of Chemistry, Center for Molecular Diversity in Drug Design, Discovery and Delivery, University of South Florida, Tampa, FL 33620, USA, bjbaker@usf.edu

Mark A. Bernards Department of Biology and the Biotron, The University of Western Ontario, London, ON N6A 5B7, Canada, bernards@uwo.ca

Paul W. Bosland Department of Plant and Environmental Sciences, New Mexico State University, Las Cruces, NM 88003, USA, pbosland@nmsu.edu

Dan Brown Agriculture and Agri-Food Canada, London, ON, Canada, dan.brown@agr.gc.ca

Cynthia J. Bucher Department of Global Health, Center for Biological Defense, University of South Florida, Tampa, FL 33620, USA, cbucher@health.usf.edu

Vincenzo De Luca Department of Biological Sciences, Brock University, St. Catharines, ON L2S 3A1, Canada, vdeluca@brocku.ca

Richard A. Dixon Plant Biology Division, Samuel Roberts Noble Foundation, Ardmore, OK 73401, USA, radixon@noble.org

Brian C. Foster Department of Cellular and Molecular Medicine, University of Ottawa, Ottawa, ON, Canada; Therapeutic Products Directorate, Health Canada, Ottawa, ON, Canada, brian_foster@hc-sc.gc.ca

Ivette Guzman Department of Plant and Environmental Sciences, New Mexico State University, Las Cruces, NM 88003, USA, iguzman@unity.ncsu.edu

Takefumi Hattori Research Institute for Sustainable Humanosphere, Kyoto University, Kyoto 611-0011, Japan, thattori@rish.kyoto-u.ac.jp

Dimitre A. Ivanov Department of Biology and the Biotron, The University of Western Ontario, London, ON N6A 5B7, Canada, divanov2@uwo.ca

Nikola Kovinich Bioproducts and Bioprocesses, Research Branch, Agriculture and Agri-Food Canada, Ottawa, ON K1A 0C6, Canada; Department of Biology, Carleton University, Ottawa-Carleton Institute of Biology, Ottawa, ON K1S 5B6, Canada, kovinichn@agr.gc.ca

Sungbeom Lee Department of Biological Sciences, Virginia Tech, Blacksburg, VA 24061, USA, sungbeom@vt.edu

E.M.K. Lui Department of Physiology and Pharmacology, University of Western Ontario, London, ON, Canada, ed.lui@schulich.uwo.ca

Alice Luu Department of Biology, McMaster University, Hamilton, ON, Canada, alice_luu86@hotmail.com

J. Alan Maschek Department of Chemistry, Center for Molecular Diversity in Drug Design, Discovery and Delivery, University of South Florida, Tampa, FL 33620, USA, jmaschek@mail.usf.edu

James B. McClintock Department of Biology, University of Alabama at Birmingham, Birmingham, AL 35294, USA, mcclinto@uab.edu

Cecilia A. McIntosh Department of Biological Sciences, East Tennessee State University, Johnson City, TN 37614, USA; School of Graduate Studies, East Tennessee State University, Johnson City, TN 37614, USA, mcintosc@etsu.edu

Kristina L. McIntyre Department of Biology, University of Ottawa, Ottawa, ON, Canada, kmcin009@uottawa.ca

Brian Miki Department of Biology, Carleton University, Ottawa-Carleton Institute of Biology, Ottawa, ON K1S 5B6, Canada, mikib@agr.gc.ca

Tomoyuki Nakatsubo Research Institute for Sustainable Humanosphere, Kyoto University, Kyoto 611-0011, Japan; Central Laboratory, Rengo Co., Ltd., 4-1-186, Ohiraki, Fukushima-ku, Osaka 553-0007, Japan, tomoyuki@hyper.ocn.ne.jp

M. Andreea Neculai Department of Biology and the Biotron, The University of Western Ontario, London, ON N6A 5B7, Canada, m.neculai@gmail.com

Robert W. Nicol University of Guelph Ridgetown Campus, Ridgetown, ON N0P 2C0, Canada, rnicol@ridgetown.uoguelph.ca

Mary A. O'Connell Department of Plant and Environmental Sciences, New Mexico State University, Las Cruces, NM 88003, USA, moconnel@nmsu.edu

Alberto van Olphen Department of Global Health, Center for Biological Defense, University of South Florida, Tampa, FL 33620, USA, avanolph@health.usf.edu

Daniel K. Owens Department of Biological Sciences, East Tennessee State University, Johnson, TN 37614, USA, owens1@etsu.edu

Giulio M. Pasinetti Department of Psychiatry, Mount Sinai Medical Center, New York, NY 10029, USA, giulio.pasinetti@mssm.edu

M. Soledade C. Pedras Department of Chemistry, University of Saskatchewan, Saskatoon, SK S7N 5C9, Canada, s.pedras@usask.ca

Ralph L. Reed Department of Pharmaceutical Sciences, Oregon State University, Corvallis, OR 97331, USA, reedr@oregonstate.edu

Vaishali Sharma Plant Biology Division, Samuel Roberts Noble Foundation, Ardmore, OK 73401, USA, vaishalisharma@gmail.com

Jan F. Stevens Department of Pharmaceutical Sciences, Oregon State University, Corvallis, OR 97331, USA, fred.stevens@oregonstate.edu

Cathy Sun Department of Biology, University of Ottawa, Ottawa, ON, Canada, cathy.j.sun@gmail.com

Shiro Suzuki Institute of Sustainability Science, Kyoto University, Kyoto 611-0011, Japan; Research Institute for Sustainable Humanosphere, Kyoto University, Kyoto, 611-0011, Japan, shiro-s@rish.kyoto-u.ac.jp

Teresa W. Tam Department of Cellular and Molecular Medicine, University of Ottawa, Ottawa, ON, Canada, teresa_w_tam@hc-sc.gc.ca

Dorothea Tholl Department of Biological Sciences, Virginia Tech, 24061, Blacksburg, VA, USA, tholl@vt.edu

Toshiaki Umezawa Research Institute for Sustainable Humanosphere, Kyoto University, Kyoto 611-0011, Japan; Institute of Sustainability Science, Kyoto University, Kyoto 611-0011, Japan, tumezawa@rish.kyoto-u.ac.jp

Masaomi Yamamura Research Institute for Sustainable Humanosphere, Kyoto University, Kyoto 611-0011, Japan, zenstyle@rish.kyoto-u.ac.jp

Chungfen Zhang Plant Biology Division, Samuel Roberts Noble Foundation, Ardmore, OK 73401, USA, zhangch@msu.edu

Chapter 1
The Pursuit of Potent Anti-influenza Activity from the Antarctic Red Marine Alga *Gigartina skottsbergii*

J. Alan Maschek, Cynthia J. Bucher, Alberto van Olphen, Charles D. Amsler, James B. McClintock, and Bill J. Baker

1.1 The Necessity for Antiviral Treatments

Humans, plants, insects, and other animals are all susceptible to viral infection; therefore, prevention and control of viral disease carry important health and economic implications. The common cold, acquired immune deficiency syndrome (AIDS), and some cancers are carried by viruses. Viral plant diseases are known to impact fruit trees, tobacco, and many vegetables [1]. Both insects and animals have the ability to transfer viral disease to humans and other animals. The health and economic consequences of viral disease carry enormous consequences, and significant advances have been made toward amelioration of antiviral threats. There is a critical need to identify novel drug classes and new chemical structures, which can be exploited for antiviral drug development.

Influenza is an enveloped RNA virus infecting birds and mammals. The World Health Organization (WHO) estimates that 3–5 million people are infected each year, and as many as 500,000 people die from the complications of influenza infections. Three influenza pandemics have occurred within the past century. The deadliest outbreak ever recorded (1918–1919) killed about 40 million people worldwide, including about 650,000 in the United States. Both seasonal and pre-pandemic influenza are a reality that requires continued effort for the development of vaccines and novel drugs. The emergence of new influenza viral strains, for which the population is immunologically naive, represents an undeniable challenge and a continued threat for the military and civilian population. The current lag in vaccine production underscores the role that anti-influenza virus drugs play in the treatment and mitigation of the spread of this viral infection. Existing treatments have significant limitations: the list of FDA-approved drugs effective against seasonal, avian, and the most recent pre-pandemic swine influenza is composed of two anti-neuraminidase compounds with the same chemical scaffold (oseltamivir and zanamivir). The M2

B.J. Baker (✉)
Department of Chemistry, Center for Molecular Diversity in Drug Design, Discovery and Delivery, University of South Florida, Tampa, FL 33620, USA
e-mail: bjbaker@usf.edu

D.R. Gang (ed.), *The Biological Activity of Phytochemicals*, Recent Advances in Phytochemistry 41, DOI 10.1007/978-1-4419-7299-6_1,
© Springer Science+Business Media, LLC 2011

ion channel blockers amantadine and rimantadine have long been in use around the globe; however, they are not well tolerated, are ineffective against the avian H5N1 and 2009 H1N1 strains, and drug-resistant strains have been isolated [2–6]. To further darken the therapeutic scenario of flu treatment, both natural and drug-induced resistances to these medicines are widespread and possible links to the appearance of neurological side effects from neuraminidase inhibitors have been reported [7–9].

1.2 Natural Products: Red Marine Algae and the Antarctic Peninsula

Dietary supplement-producing companies are currently marketing a *Gigartina*-based red marine algal (RMA) extract as a "flu fighter," with claims of sulfated polysaccharides that boost immune response and fight influenza, although they acknowledge these claims have not been evaluated by the Food and Drug Administration (FDA) [10]. RMA are known to contain significant amounts of sulfated polysaccharides, also known as carrageenans, which have demonstrated a broad range of biological activities, most notably against human immunodeficiency virus (HIV), Herpes simplex virus (HSV) types 1 and 2, and respiratory syncytial virus (RSV) [11], as well as variable effects on the immune system and anticoagulant activity [12, 13].

Gigartina is widely cultivated as a source of carrageenans for the food additive industry. Carrageenans are a family of linear sulfated polysaccharides found in many red algae. Various species, including *Gigartina stellata*, *G. chamissoi*, *G. tenella*, and *G. skottsbergii*, are farmed for commercial use, the majority cultivated from suspended rafts along the coasts of Chile. These carbohydrates are gelatinous and stable under stress and have been used in the food industry as thickening and stabilizing agents. Other products where carrageenans are used are as follows: desserts, ice cream, beer, toothpaste, various gels, shampoo, marbling, shoe polish, biotechnology, diet soda, soy milk, pet food, personal lubricants, and as a health supplement.

In preparation for this project, extensive collecting of marine invertebrates, algae, and microorganisms was performed from the shallow water environment of the western Antarctic Peninsula, where recent efforts have revealed a diverse fauna replete with defensive chemistry [14]. Screening of extracts from these organisms against influenza viruses has revealed that the red marine alga *G. skottsbergii* elaborates potent anti-influenza chemistry. It is of note that although this alga in Antarctica is morphologically very similar to *G. skottsbergii* from Chile, recent molecular taxonomic studies indicate that it is probably a distinct species [15, 16]. Nevertheless, until a new species designation is formally proposed and published, *G. skottsbergii* remains the correct name for this entity. This perspective describes the bioassay-guided fractionation utilizing a cell-based primary screening assay to rapidly identify natural product extracts exhibiting inhibitory activity on viral growth evidenced by reduction in the development of cytopathic effect (CPE) on susceptible cultured cells. A hemagglutination assay has provided some insight into the possible mechanisms of action by interference of viral docking. SDS-PAGE gel electrophoresis and sequencing of the active fraction reveal some homology with actin and/or lectins.

1.3 Primary and Secondary Screening of Extract Library

Initial screening of lipophilic and hydrophilic extracts against influenza A/WY/03/2003 (H3N2) was selected from a library of diverse marine invertebrates, algae, and microorganisms. The primary influenza screen used in this study begins with a microscopic evaluation of the cytopathic effect of extracts on virus-infected mammalian cells and is quantified by an MTT stain. From 800 screened extracts, only one, well A4 in Fig. 1.1, which is the crude extract from *G.*

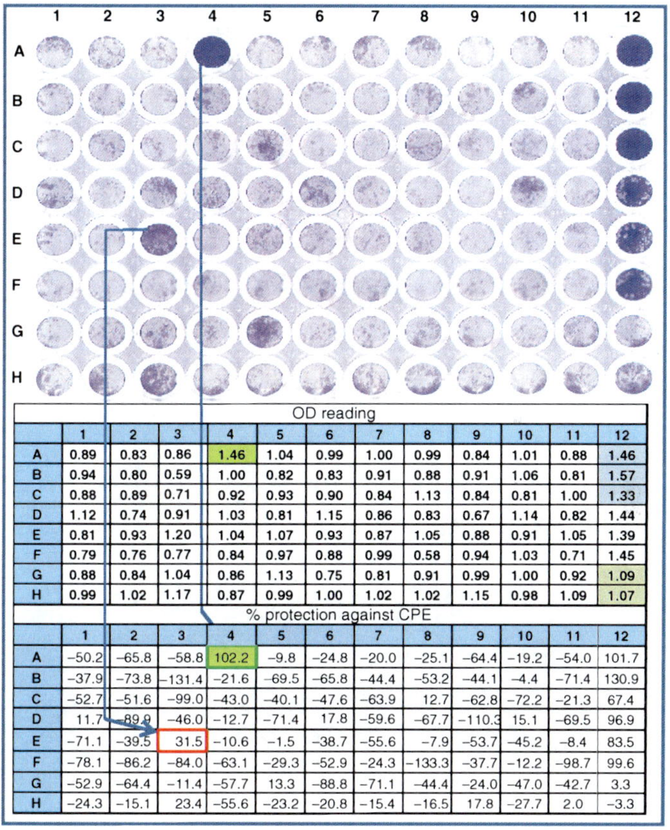

OD reading											
	1	2	3	4	5	6	7	8	9	10	11	12
A	0.89	0.83	0.86	1.46	1.04	0.99	1.00	0.99	0.84	1.01	0.88	1.46
B	0.94	0.80	0.59	1.00	0.82	0.83	0.91	0.88	0.91	1.06	0.81	1.57
C	0.88	0.89	0.71	0.92	0.93	0.90	0.84	1.13	0.84	0.81	1.00	1.33
D	1.12	0.74	0.91	1.03	0.81	1.15	0.86	0.83	0.67	1.14	0.82	1.44
E	0.81	0.93	1.20	1.04	1.07	0.93	0.87	1.05	0.88	0.91	1.05	1.39
F	0.79	0.76	0.77	0.84	0.97	0.88	0.99	0.58	0.94	1.03	0.71	1.45
G	0.88	0.84	1.04	0.86	1.13	0.75	0.81	0.91	0.99	1.00	0.92	1.09
H	0.99	1.02	1.17	0.87	0.99	1.00	1.02	1.02	1.15	0.98	1.09	1.07

% protection against CPE											
	1	2	3	4	5	6	7	8	9	10	11	12
A	−50.2	−65.8	−58.8	102.2	−9.8	−24.8	−20.0	−25.1	−64.4	−19.2	−54.0	101.7
B	−37.9	−73.8	−131.4	−21.6	−69.5	−65.8	−44.4	−53.2	−44.1	−4.4	−71.4	130.9
C	−52.7	−51.6	−99.0	−43.0	−40.1	−47.6	−63.9	12.7	−62.8	−72.2	−21.3	67.4
D	11.7	−89.9	−46.0	−12.7	−71.4	17.8	−59.6	−67.7	−110.3	15.1	−69.5	96.9
E	−71.1	−39.5	31.5	−10.6	−1.5	−38.7	−55.6	−7.9	−53.7	−45.2	−8.4	83.5
F	−78.1	−86.2	−84.0	−63.1	−29.3	−52.9	−24.3	−133.3	−37.7	−12.2	−98.7	99.6
G	−52.9	−64.4	−11.4	−57.7	13.3	−88.8	−71.1	−44.4	−24.0	−47.0	−42.7	3.3
H	−24.3	−15.1	23.4	−55.6	−23.2	−20.8	−15.4	−16.5	17.8	−27.7	2.0	−3.3

Fig. 1.1   Representative of primary screening results for a library of marine extracts. Crystal violet-stained plate (*top panel*), optical density (OD reading) and corresponding percentage protection (% protection against CPE) deduced from the cell viability assay. The crude extract in position A4 (*Gigartina skottsbergii*) was determined to protect against influenza-induced CPE using light microscopy and cell viability assay at 100 μg/mL. The extract in position E3 partially protected against CPE. Wells A-C12 contained uninfected cell control, D-F12 contained control drug ribavirin at 5 μg/mL, and G12 and H12 are virus-infected control

skottsbergii, displayed complete protection of the mammalian host cells. For comparison, healthy, unaffected cells in this assay stain violet, as shown in both well A4, and along the control lane, column 12.

Extracts that exhibited significant inhibitory activity, defined as $\geq$50% inhibition of CPE at 100 μg/mL, were advanced into secondary screening, which includes confirmation of activity observed during the primary screen using an expanded range of concentrations (dose response), plaque inhibition assay, and a one-step growth inhibition and testing of additional influenza viruses. As shown in Fig. 1.2a, the quantitative dose response assay was used to assess the potency of the most two

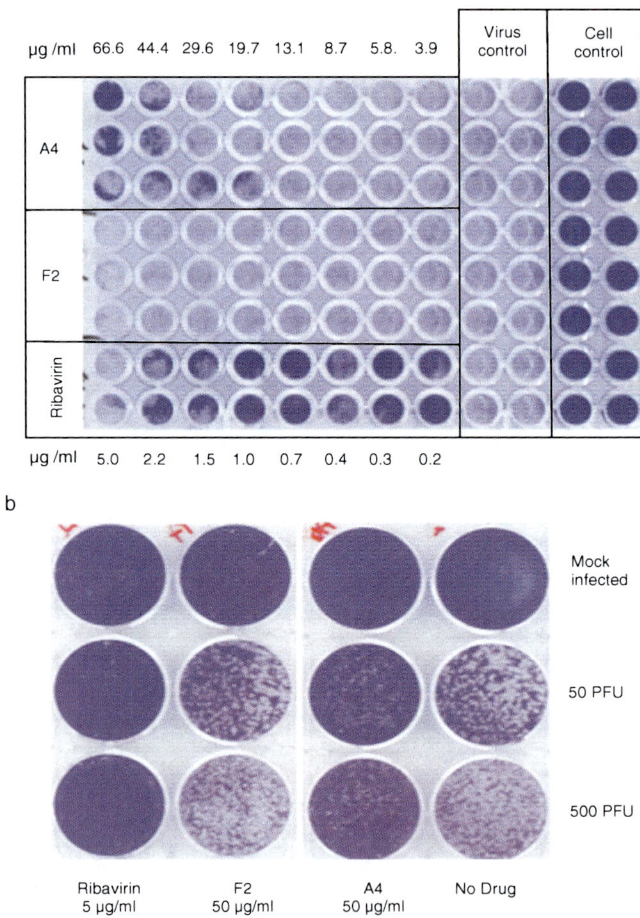

Fig. 1.2 **a** Microtiter plate illustrating dose response activity of an extract of the red alga *Gigartina skottsbergii* (A4) as well as unrelated and ineffective extract in well F2. **b** Plaque reduction assay depicting a marked reduction of virus replication in cells treated with extract A4 (50 μg/mL) compared with "No Drug" or the unrelated ineffective extract F2. The effect of A4 is similar to that of the control drug ribavirin

active fractions, A4 and F2. Extracts were serially diluted 2/3-folds over eight concentrations and the percentage protection determined using the same approach as the one described for the primary screening. The extract in well A4 displayed activity against H3N2 influenza virus in vitro in a dose-dependent manner as visualized by the decreasing amounts of violet stain going across the well plate. A plaque reduction assay was also performed, demonstrating the extract present in well A4 performed almost as well as in the control drug ribavirin (Fig. 1.2b). Briefly, a plaque assay is a method whereby infected cells form plaques – shown here as the circular holes in the stain allowing quantification of the effectiveness of the drug by looking at both the size and number of the plaques formed. The active extract A4 induced the formation of smaller plaques than the untreated or extract controls.

Additional assays included a selectivity evaluation and cytotoxicity assay where both good selectivity and low cytotoxicity were documented. The extract was also tested on unrelated viruses and other influenza viruses. Extract A4 was the only one to maintain the activity through secondary screening, and with that as a target, purification of the active extract commenced.

1.4 Purification of Active Extract

The thawed alga was vigorously extracted in methanol and then partitioned between organic solvents and water to derive five fractions, of which all three of the organic fractions exhibited antiviral activity (Fig. 1.3a). Fraction A performed the best, possessing an EC_{50} value of 39 μg/mL. Subfraction I – a very polar fraction – showed the strongest activity, with an EC_{50} value of 28 μg/mL. Reverse phase HPLC evolved 11 fractions. However, when these fractions were submitted for bioassay, activity had been lost – even when the fractions were recombined. Traditional spectroscopic techniques did not provide insight into the nature of the active subfraction I, leading to the tentative conclusion that the extract did not contain small-molecule (i.e., natural product) components. Concentration of the active fractions with high molecular weight filters and SDS-PAGE gel electrophoresis provided evidence that the active fraction could be a protein or a polymeric. Sequencing data from the active fraction, although not yet conclusive, exhibited homology with actins and/or lectins at $\leq 85\%$ confidence levels. Because of the failure to evolve further active fractions from traditional small molecule purification schemes, chromatographic methodology for proteins was undertaken. As shown in Fig. 1.3b, thawed algae were blended and extracted in water and then adsorbed onto XAD-1180 resin multiple times to eventually yield four fractions. Fraction 3 that eluted with 1 M NaOH – the fraction the manufacturer's protocol recommended for the elution of proteins and peptides – was active in the assay. A size exclusion column separation of fraction 3 generated 10 fractions. Subfraction A was active toward both the Wyoming H3N2 virus and H1N1, with EC_{50} values of 30 and 10 μg/mL, respectively. Although activity is reproducible among the disparate separation methods, a pure, active compound has not yet been obtained.

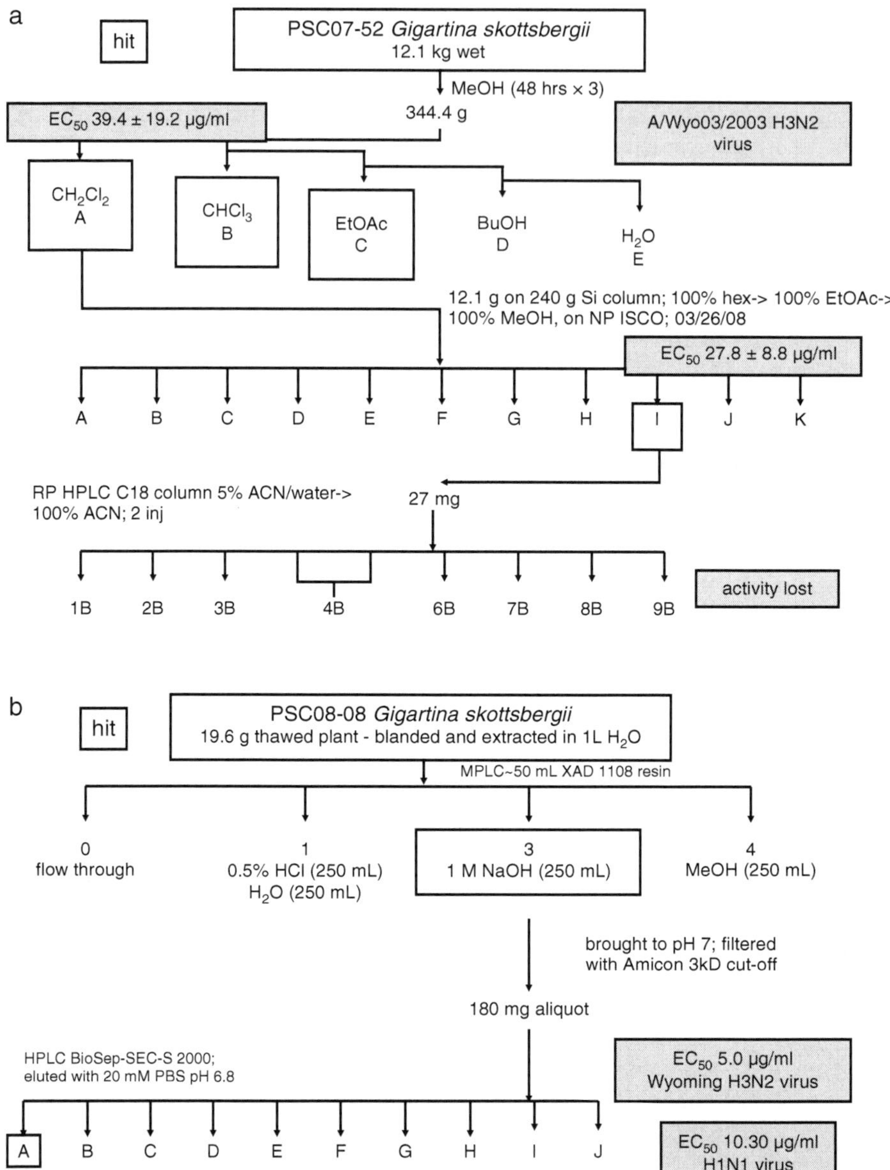

Fig. 1.3 Fractionation schemes for extracts from the red alga *Gigartina skottsbergii*. **a** The initial classic small molecule isolation scheme and **b** alternative protein isolation scheme

1.5 Insight into the Mechanism of Action via Hemagglutination Assay and Drug Combination Studies

Hemagglutination assays (HAs) have provided some insight into the possible mechanism of action. Hemagglutination is the agglutination involving red blood cells whereby proteins on the surface of the virus bind – or hemagglutinate – to the neuraminic acid receptors allowing for quantification of viral dilution and infection. A positive result is denoted in this assay by visible agglutination of the red blood cells when compared with the negative control well, observed as the settling of the red blood cells in the test well. When neuraminic acid receptor mimics are introduced, hemagglutination interference occurs and the erythrocytes will settle to the bottom of the well. When neuraminic acid binding compounds are introduced, hemagglutination will occur with protection of the cells as the mimics are outcompeting the surface of the erythrocyte. As shown in Fig. 1.4a, hemagglutination occurs when untreated cells are infected with virus (Virus A/WSN33), resulting in a suspension of the virus and erythrocyte binding in solution. Cells treated with the active fraction (Extract I + Virus) exhibit agglutination as well, whereby the extract is protecting the uninfected cell from virus by blocking viral entry. Rows "Extract I only" and "Virus

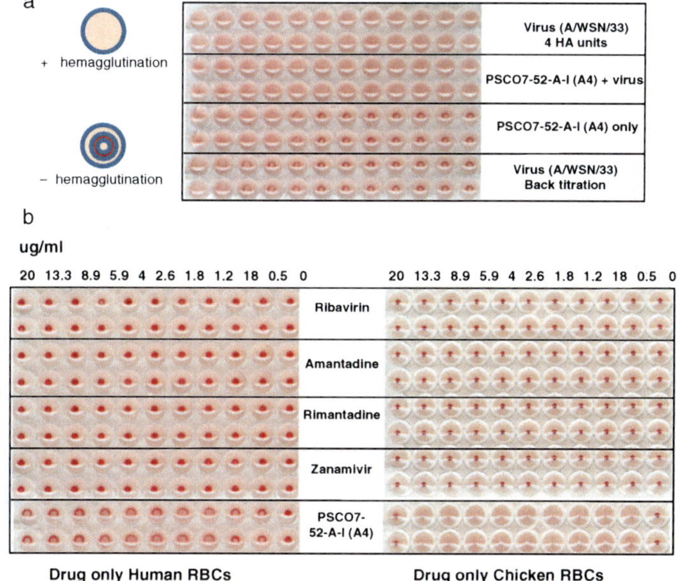

Fig. 1.4 Hemagglutination assay results from the active fraction of the red alga *Gigartina skottsbergii*. **a** Red blood cells (RBCs) hemagglutinate in the presence of influenza virus (*top* row) and with extract A4 + virus or extract A4 alone. The *bottom* two rows present the back titration of the virus used in this experiment. **b** In contrast to the other anti-influenza drugs (ribavirin, amantadine, rimantadine, and zanamivir) that do not induce hemagglutination of human as well as chicken RBCs, extract A4 does it in a dose-dependent manner

HA" are controls used for quantification of the assay. As shown in Fig. 1.4b, the control drugs were observed causing no hemagglutination while protecting the cell. Subfraction I, however, protects the cell while exhibiting a complete to almost complete hemagglutination in the well. This suggests that a protein in subfraction I may be competing with the viral surface proteins and working by a different mechanism than control drugs.

With the noted differences observed between the control drugs and subfraction I in the assays, drug combination assays were conducted. The results revealed impressive results as combinations with the active subfraction I (Fig. 1.3), and the drugs zanamivir and ribavirin consistently observed protection greater than the sum of each. Zanamivir (2.5 μg/mL) displayed 10.4% protection, the active subfraction I (8.8 μg/mL) exhibited 41.1% protection, but together proved to protect 59.1% of the cells from infection (Fig. 1.5a). Experiments with ribavirin showed even greater increased response. Ribavirin (1.25 μg/mL) and subfraction I (8.8 μg/mL)

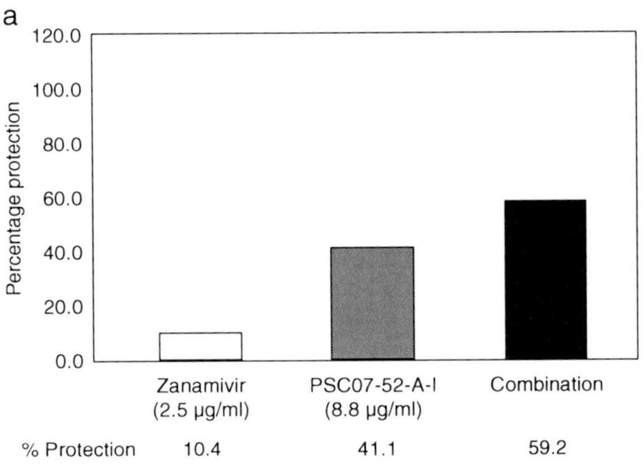

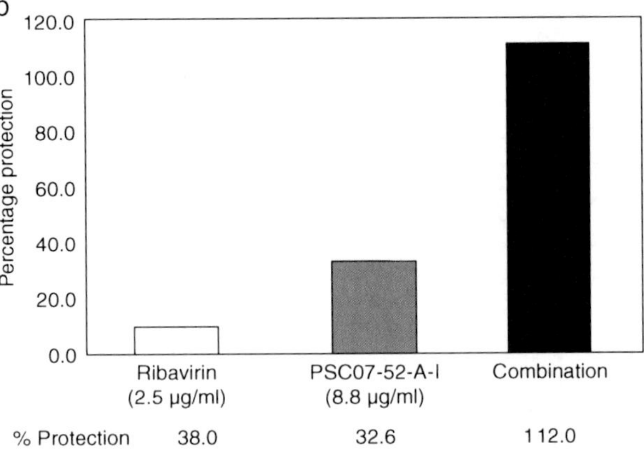

Fig. 1.5 Percentage protection results for combination drug assays using extract of the red alga *Gigartina skottsbergii* PSC07-52-A-I (A4) at 8.8 μg/mL with (**a**) zanamivir at 2.5 μg/mL alone and in combination with PSC07-52-A-I and (**b**) ribavirin at 2.5 μg/mL alone and in combination with PSC07-52-A-I

displayed 38.0 and 32.6% protection, respectively (Fig. 1.5b), but in combination exhibited complete protection, suggesting a synergistic effect and perhaps offering potential value in drug combination therapies.

1.6 Conclusions

These results emphasize the importance of exploring the marine realm for the presence of antiviral compounds. Mori et al. [17] have reported a potent antiviral lectin from the red alga *Griffithsia* sp., which is perhaps similar in nature to the active component of *G. skottsbergii*. Subfraction I possesses strong anti-influenza activity toward both the Wyoming H3N2 virus and H1N1 virus, with EC_{50} values in the range of 5–10 μg/mL. This chapter validates a successful cell-based screening assay capable of accurately identifying active crude extracts from extensive collections. In addition, this in vitro assay was able to reproduce the published EC_{50} values of known antivirals (i.e., ribavirin, amantadine, rimantadine, and zanamivir). These results underscore the value of screening natural products not only for identification of small molecules but also as a source of potent macromolecules with antiviral activity.

1.7 Experimental

General Experimental Procedures: MPLC fractionation was performed using an Isco CombiFlash, and HPLC isolation was performed using Shimadzu pumps and detector and YMC-Pack ODS-AQ C18 column.

Plant Material: Collection of the rhodophyte *G. skottsbergii* was done manually during scuba dives within 3.5 km of Palmer Station on Anvers Island off the western Antarctic Peninsula (64° 46.5' S, 64° 03.3' W) at a depth of 5–12 m. Identifications were made by Prof. Bill J. Baker (University of South Florida) and Prof. Charles D. Amsler (University of Alabama at Birmingham).

Extraction and Isolation: Thawed algae (12.1 kg) were extracted with MeOH (4 L × 3). The combined extract was concentrated (344.4 g), and the residue was partitioned between CH_2Cl_2 and H_2O. Subsequently, the organic layer was concentrated in vacuo to give a dark green crude (20.2 g). The residue was subjected to Si gel column chromatography with a hexanes/EtOAc/MeOH gradient solvent system to give 12 fractions. The active fraction (2.9 g) eluted with approximately 50–75% MeOH/EtOAc and then fractionated by RP HPLC (YMC-PAK ODS-AQ), with a gradient of 50% aqueous MeOH to 100% MeOH to yield eight fractions. The active fraction (11 mg) eluted in approximately 75% $MeOH/H_2O$. Whole algae thalli (200 g (wet wt.)) were used for the preparation of protein extracts using the commercial kit P-PER (Pierce). Using membrane filtration, the aqueous phase was sequentially concentrated by Amicon filter devices of decreasing size (30, 10, and 3 kDa), evolving three retentate and three filtrate fractions. Retentate 1 (MW ≥30 kDa) possessed the strongest antiviral activity and decreased proportionally in subsequent fractions. There was no activity in the final filtrate (MW <3 kDa). Thawed algae (400 g) was blended and extracted in H_2O; after concentration the light red gel was loaded onto XAD-1180 resin and eluted with 0.5% HCl, 1 M NaOH, and 100% MeOH to evolve three fractions. The basic fraction was expected to elute peptides and proteins, and subsequent bioassay identified this fraction as active.

Thawed algae (180 g) were blended and extracted in H_2O (600 mL × 4) and then lyophilized. The pink solid was brought up to a concentration of 25 mg/mL in H_2O and then put on ice in plastic conical centrifuge tubes (15 mL × 2). Cold EtOH of 15 mL was added to each tube and allowed to settle on ice for 1 h and then kept overnight at 4°C. The tubes were centrifuged at 3,000 rpm for 1 h and supernatants combined and subsequently reduced by rotavap and lyophilizer. The lyophilized supernatant was brought to a concentration of 25 mg/mL in ddH_2O and put on ice. Crystalline molecular biology grade NH_4SO_4 (19 g) was added to bring the percentage saturation to 75%. The supernatant was allowed to precipitate on ice for 2 h and then centrifuged at 3,000 rpm for 45 min. The supernatant was poured off, and the pellet was resuspended in a minimum amount of sterile PBS (1.8 mL) with 0.02% sodium azide added to prevent bacterial growth. The amount of protein was quantitated via NanoDrop to give a concentration of 2.18 mg/mL.

Cells and Viruses: Cell cultures used in screening are Madin-Darby canine kidney (MDCK) cells, obtained from American Type Culture Collection (Manassas, VA, CCL-34, passage 55) and grown in Eagle's minimum essential medium (MEM, Invitrogen) with 10% reconstituted fetal calf serum (HyClone III). The cells are trypsinized and then resuspended at 3×10^5 cells per mL in assay media (for secondary screening DMEM, high glucose without phenol red) supplemented with gentamicin and 0.5% BSA for all subsequent steps. Cells are plated manually and incubated at 37°C and 5.0% CO_2 for 24 h prior to virus addition. Influenza virus stocks are prepared by growing influenza strains A/PR8/38 (H1N1), A/Wyoming/3/2003 (H2N3), and B/Lee/40 in MDCK cells. The supernatant from infected MDCK cells is serially diluted and used for isolation of a single plaque. A single plaque from the second round of plaque purification is selected and resuspended in serum-free Dulbecco's modified Eagle's medium (DMEM, Invitrogen, Carlsbad, CA) containing 0.35% bovine serum albumin (BSA, Invitrogen, Fraction V). The plaque-purified virus is used to inoculate three T-150 flasks containing MDCK cells (see below) at a multiplicity of infection of 0.001 PFU per cell. The supernatant is collected 72 h post infection, aliquotted, and stored a −80°C until needed.

Primary Screening: A single-dose (100 μg/mL), single-well per extract was tested in 96-well plates. Reduction of CPE was qualitatively evaluated by direct observation of cytopathic effect using an inverted light microscope. Optical density was measured at absorbance of 490 using a BioTek HT plate reader. The percentage of protection was calculated using the following formula: $(1-((\mu c-OD$ of sample$)/(\mu c-\mu v)))\times 100$, where μc is the mean optical density (OD) value of the uninfected cells and μv the mean OD value of the infected cells. After measurement of the cell viability, the plates were stained using a 2.5% crystal violet solution in PBS containing 4% formaldehyde.

Secondary Screening: A plaque reduction assay was used to test the ability of the extract to inhibit the growth of the virus in multiple rounds of replication. For these experiments, six-well plates containing 80% confluent MDCK cells monolayer were inoculated with the indicated PFU and incubated for 1 h at 37°C before adding a semisolid agar overlay containing 50 μg/mL of extract. Ribavirin was used as drug control at 5 μg/mL. The plates were incubated at 37°C for 72 h and then stained using crystal violet/formalin solution. An uninfected well containing 50 μg/mL of the extracts was included to evaluate the toxicity of the compounds. Ribavirin completely inhibited the formation of plaques at a dose of 5 μg/mL.

Dose-Dependent Response: Extracts are serially diluted 2/3-fold over eight different concentrations (Fig. 1.2) using a Biomek 3000 (Beckman, Fullerton, CA), and the percentage protection is determined using the same approach as the one described for primary screening. Quantitative analysis is performed using the cell viability assay previously described (CellTiter 96 AQueous One Solution, Promega, Madison, WI).

Concentration response data are analyzed by a nonlinear regression logistic dose response model. Each of the crude extracts are fractionated into 10–15 subfractions. For each crude extract and the associated subfractions, IC_{50}s and IC_{90}s are determined as outlined above. All crude and subfractions of a particular marine organism are assayed simultaneously (within one assay) and include ribavirin as reference drugs using only A/WY/03/2003 virus.

Drug Combination Assay: In a 96-well assay plate, extracts are serially diluted 2/3-fold over nine different concentrations from columns one through nine as above. Zanamivir or ribavirin was serial diluted 0.5-fold in duplicate over three concentrations in rows A through F. The remaining wells were used for extracts and compounds uncombined, and the negative and positive control. Cells and virus are added as previously described. The percentage protection is determined using the same approach as the one described for the primary screening. The quantitative analysis is performed using the cell viability assay previously described (CellTiter 96 AQueous One Solution, Promega). Concentration response data are analyzed by a nonlinear regression logistic dose response model.

Acknowledgments The assistance of Rachelle Price, who conducted the hemagglutination assays reported here, is gratefully acknowledged. Funding in support of this work was received from FCOE-BITT (GALS-006 to AvO and BJB) and the Office of Polar Programs of the National Science Foundation (OPP-0442769 to CDA and JBM, OPP-0442857 to BJB). Field work was facilitated by the staff of the Antarctic Support Services of the Office of Polar Programs and Raytheon Polar Services.

References

1. Pappu HR, Jones RAC, Jain RK (2009) Global status of tospovirus epidemics in diverse cropping systems: successes achieved and challenges ahead. Virus Res 141:219–236
2. Bright RA, Medina M-j, Xu X, Perez-Oronoz G, Wallis TR, Davis XM, Povinelli L, Cox NJ, Klimov AI (2005) Incidence of adamantane resistance among influenza A (H3N2) viruses isolated worldwide from 1994 to 2005: a cause for concern. Lancet 366:1175–1181

3. Bright RA, Shay DK, Shu B, Cox NJ, Klimov AI (2006) Adamantane resistance among influenza a viruses isolated early during the 2005–2006 influenza season in the United States. JAMA 295:891–894

4. Hall M, Brown Michael D (2005) Evidence-based emergency medicine/systematic review abstract. Are amantadine and rimantadine effective in healthy adults with acute influenza? Ann Emerg Med 46:292–293

5. Keyser LA, Karl M, Nafziger AN, Bertino JS Jr. (2000) Comparison of central nervous system adverse effects of amantadine and rimantadine used as sequential prophylaxis of influenza a in elderly nursing home patients. Arch Intern Med 160:1485–1488

6. Stange KC, Little DW, Blatnik B (1991) Adverse reactions to amantadine prophylaxis of influenza in a retirement home. J Am Geriatr Soc 39:700–705

7. Chotpitayasunondh T, Ungchusak K, Hanshaoworakul W, Chunsuthiwat S, Sawanpanyalert P, Kijphati R, Lochindarat S, Srisan P, Suwan P, Osotthanakorn Y, Anantasetagoon T, Kanjanawasri S, Tanupattarachai S, Weerakul J, Chaiwirattana R, Maneerattanaporn M, Poolsavathitikool R, Chokephaibulkit K, Apisarnthanarak A, Dowell Scott F (2005) Human disease from influenza A (H5N1), Thailand, 2004. Emerg Infect Dis 11:201–209

8. Nicholson KG, Wood JM, Zambon M (2003) Influenza. Lancet 362:1733–1745

9. Yen H-L, Herlocher LM, Hoffmann E, Matrosovich MN, Monto AS, Webster RG, Govorkova EA (2005) Neuraminidase inhibitor-resistant influenza viruses may differ substantially in fitness and transmissibility. Antimicrob Agents Chemother 49:4075–4084

10. Butelli E, Titta L, Giorgio M, Mock HP, Matros A, Peterek S, Schijlen EGWM, Hall RD, Bovy AG, Luo J, Martin C (2008) Enrichment of tomato fruit with health-promoting anthocyanins by expression of select transcription factors. Nat Biotechnol 26:1301–1308

11. Smit AJ (2004) Medicinal and pharmaceutical uses of seaweed natural products: a review. J Appl Phycol 16:245–262

12. Carlucci MJ, Pujol CA, Ciancia M, Noseda MD, Matulewicz MC, Damonte EB, Cerezo AS (1997) Antiherpetic and anticoagulant properties of carrageenans from the red seaweed *Gigartina skottsbergii* and their cyclized derivatives: correlation between structure and biological activity. Int J Biol Macromol 20:97–105

13. Pujol CA, Scolaro LA, Ciancia M, Matulewicz MC, Cerezo AS, Damonte EB (2006) Antiviral activity of a carrageenan from *Gigartina skottsbergii* against intraperitoneal murine *Herpes simplex* virus infection. Planta Med 72:121–125

14. Lebar MD, Heimbegner JL, Baker BJ (2007) Cold-water marine natural products. Nat Prod Rep 24:774–797

15. Hommersand MH, Fredericq S (2003) Biogeography of the marine red algae of the South African West Coast: a molecular approach. In: Chapman, RO, RJ Anderson, VJ, Vreeland and IR Davison (eds) Proceedings 28th International Seaweed Symposium. Oxford University Press, Oxford, pp. 325–336

16. Hommersand MH, Moe RL, Amsler CD, Fredericq S (2009) Notes on the systematics and biogeographical relationships of Antarctic and sub-Antarctic Rhodophyta with descriptions of four new genera and five new species. Bot Mar 52:509–534

17. Mori T, O'Keefe BR, Sowder RC II, Bringans S, Gardella R, Berg S, Cochran P, Turpin JA, Buckheit RW Jr., McMahon JB, Boyd MR (2005) Isolation and characterization of griffithsin, a novel HIV-inactivating protein, from the red alga *Griffithsia* sp. J Biol Chem 280:9345–9353

Chapter 2
Ginsenosides: Phytoanticipins or Host Recognition Factors?

Mark A. Bernards, Dimitre A. Ivanov, M. Andreea Neculai, and Robert W. Nicol

2.1 Introduction

Plants produce a wide range of natural products, also referred to as secondary metabolites, including alkaloids, isoprenoids, and phenolics, many of which possess potent biological activity. While it is often difficult to distinguish between "primary" and "secondary" metabolism, a functional definition describes secondary metabolites as those molecules that are involved in the interaction between plants and their environment [1]. From a human perspective, the biological activity of plant natural products has been exploited by their use as medicinal compounds (either directly or as starting point for the development of synthetic analogs); several examples are provided within this volume. However, plants do not make natural products for the benefit of humans, but instead as protective measures against the myriad challenges (both biotic and abiotic) that they face in their environment. The evolutionary development of natural products has been driven by these challenges [2]. In this chapter, we provide a summary of our work investigating the potential role of ginsenosides (a subclass of triterpene saponins produced by *Panax* spp.) in the ecology of *Panax quinquefolius* (American ginseng). In doing so, we draw upon the vast literature on saponins in general and provide our perspective on the ecological relevance of ginsenosides in the ginseng-*P. irregulare* pathosystem.

2.2 Saponins

2.2.1 Chemical Structure and Diversity

Saponins are glycosylated natural products with soap-like properties [3]. The aglycones (sapogenins) are either triterpenoid- or steroid-derived. Dicotyledonous

M.A. Bernards (✉)
Department of Biology and the Biotron, The University of Western Ontario, London, ON N6A 5B7, Canada
e-mail: bernards@uwo.ca

D.R. Gang (ed.), *The Biological Activity of Phytochemicals*, Recent Advances in Phytochemistry 41, DOI 10.1007/978-1-4419-7299-6_2,
© Springer Science+Business Media, LLC 2011

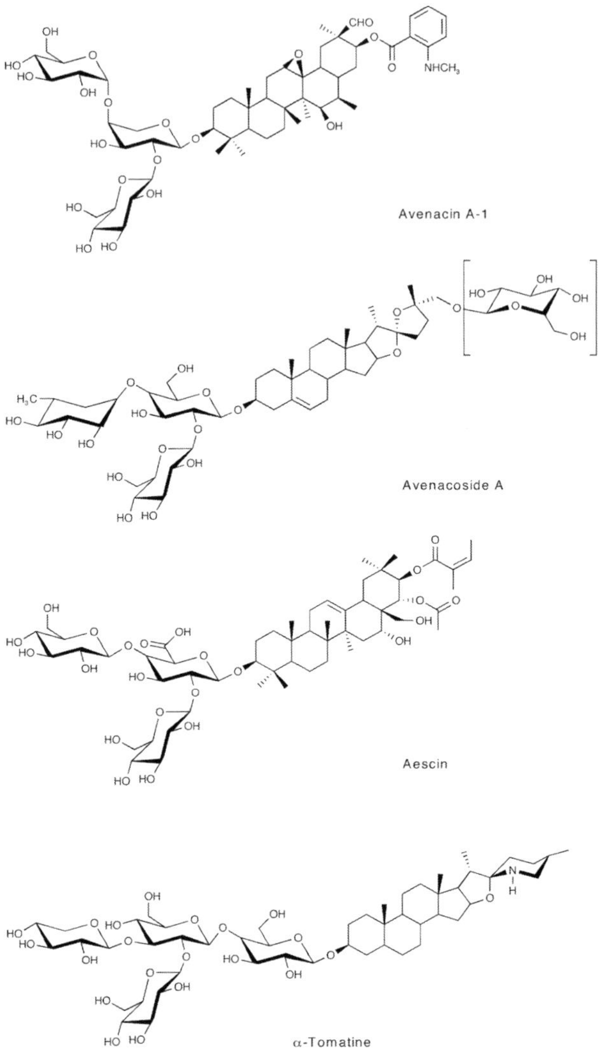

Fig. 2.1 Structures of some common saponins. The structures of some of the saponins described in the text are shown, including aescin from horse chesnut, avenacin A-1 and avenacoside A from oat, and α-tomatine from tomato. The glucose molecule enclosed in square brackets in the structure of avenacoside A highlights the glucose moiety that is cleaved off by hydrolysis by glycosidases in disrupted oat leaf tissue, leading to the fungitoxic 26-desglucosyl avenacoside A. Redrawn from [94]

plants, for example, have pentacyclic (e.g., α- and β-amyrin) or tetracyclic (e.g., the dammaranes) triterpenoid sapogenins, although in the Solanaceae, steroidal glycoalkaloid saponins (nitrogen analogs of steroidal saponins) are also present. In monocotyledonous plants, steroidal saponins are common. Examples of several common saponins are shown in Fig. 2.1.

The structural diversity of saponins is further enhanced by the number and placement of sugars around the aglycone. That is, saponins can be either mono- or bidesmosidic, with the latter having sugar moieties attached at either end of the parent carbon skeleton. Monodesmosidic saponins are generally glycosylated at the C-3 hydroxyl group; bidesmosidic saponins are generally glycosylated at the C-3 hydroxyl group as well as elsewhere in the molecule. Glucose, arabinose, galactose, rhamnose, and xylose, as well as sugar acids, may be present, either as monosaccharides or as higher order oligomers. Subtle differences in the sugar substitution pattern of a common carbon skeleton result in large numbers of compounds with closely related structures and properties.

2.2.2 Involvement of Saponins in Host–Pathogen Interactions

Saponins are biologically active phytochemicals that act upon diverse organisms, including fungi [4, 5]. For example, alfalfa, clover, oat, soybean, potato, ginseng, tomato, and pepper all contain either triterpenoid, steroid, or glycoalkaloid saponins, and many of these saponins possess fungitoxic activity. However, assessment of fungitoxicity through the use of in vitro bioassays has limited use when the goal is to elucidate host-pathogen interactions. The Osbourn group in the United Kingdom has used a variety of approaches over the past 15 years to clearly demonstrate that saponins can act as constitutive plant defenses (i.e., phytoanticipins) [6]. This study complements earlier avenacin bioassay observations [7]. Taken as a whole, the avenacin model appears to have confirmed the long-held assumption that the ecological function of saponins is to defend the plant against fungal pathogens, and this concept has taken hold within the literature [8, 9]. With acknowledgment that the "saponins as defense" story has been told by many groups using different pathosystems, the most complete story is represented by avenacin A-1 from oat and α-tomatine from tomato, which is described in the sections that follow.

2.2.2.1 Avenacin A-1 and α-Tomatine as Models for Plant Defense

Oat produces structurally diverse triterpenoid saponins in different plant organs. Bidesmosidic saponins (avenacosides) are produced in oat leaves and monodesmosidic saponins (avenacins) are produced in oat roots [6]. Avenacin A-1 is considered the main oat root saponin and is represented by a pentacyclic oleanane skeleton with a trisaccharide (arabinose and two glucose moieties) at C-3 (Fig. 2.1). Research with saponin-deficient oat mutants clearly demonstrated the link between in planta saponin content and defense against fungal pathogens. Avenacin fluorescence was used to screen for saponin-deficient (sad) mutants of a wild oat, *Avena strigosa*. Saponin-deficient mutants were found to have an increased susceptibility to the fungus *Gaeumannomyces graminis* var *tritici* (which is usually only pathogenic to wheat) as well as pathogenic species of *Fusarium* [10].

Glycoalkaloid saponins are found mainly in the Solanaceae. In tomato, for example, the glycoalkaloid saponin, α-tomatine, is found in relatively high concentrations in unripe fruits and is fungitoxic [11–13]. However, the tomato pathogen *Fusarium oxysporum* f. sp. *lycopersici* can degrade tomatine via the tomatinase enzyme Tom1. Targeted disruption and overexpression studies of the *Tom1* gene demonstrated that this enzyme is required for the pathogen to reach full virulence on this plant [14]. Heterologous expression of a tomatinase in yeast resulted in improved yeast saponin resistance [15] and provides further evidence that this glycoalkaloid saponin likely functions in plant defense.

2.2.2.2 Mode of Action Through Membrane Disruption

Avenacin A-1, α-tomatine, and other saponins are toxic to fungi because saponins form complexes with sterols [16–18] and thereby disrupt the integrity of biological membranes. Formation of pores and leakage of cell contents have been observed in membranes treated with avenacin and tomatine alike [11, 12, 16, 18]. For example, incubating fungal mycelium with α-tomatine caused loss of electrolytes from different fungal isolates. Of particular note, three pathogens of tomato were more resistant to the effects of α-tomatine than four nonpathogens [11]. It is presumed that saponins exert their effect in fungi by binding to ergosterol, the major membrane sterol of higher fungi [19].

The structure of saponins, especially the position of the sugar moieties at one (monodesmosidic) or two (bidesmosidic) sites on the sapogenin, has a great influence on fungitoxicity. Monodesmosidic, amphiphilic saponins are more fungitoxic than bidesmosidic saponins, and removal of one, some, or all of the sugars of the monodesmoside generally results in a reduction in toxicity [9, 12, 20]. A recent report [12] demonstrated that some sapogenins exerted fungitoxic effects on yeast, but it was concluded that the mode of action was not through the disruption of membranes. In stark contrast, oomycete pathogens are reported to be unaffected by saponins in vitro. It is probably the lack of ergosterol in these species that allows them to avoid saponin toxicity and provides "innate resistance" [21].

2.3 Ginsenosides Are Bidesmosidic Saponins with Mild Fungitoxicity

2.3.1 Structure, Nomenclature, and Biosynthesis of Ginseng Saponins

Although numerous secondary metabolites can be found in ginseng, many of the pharmacological virtues of this plant are generally thought to be due to triterpenoid saponins called ginsenosides [22]. Both American (*P. quinquefolius*) and Asian species of ginseng (e.g., *Panax ginseng*, *P. notoginseng*, *P. vietnamensis*) contain numerous, and often overlapping, types of ginsenosides. While there are >25 known ginsenosides [23] in American ginseng, the most common are ginsenosides Rg_1, Re, Rb_1, Rb_2, Rc, and Rd (named on the basis of their original chromatographic elution order). All ginsenosides are based on a tetracyclic dammarane skeleton [24] (Fig. 2.2), and since they have saccharides attached at both C-3 (or C-6) and C-20, they are bidesmosidic. They are further classified into two groups, according to the hydroxylation pattern of the parent aglycone and the attachment position of the various saccharide molecules. For example, ginsenosides Rb_1, Rb_2, Rc, and Rd

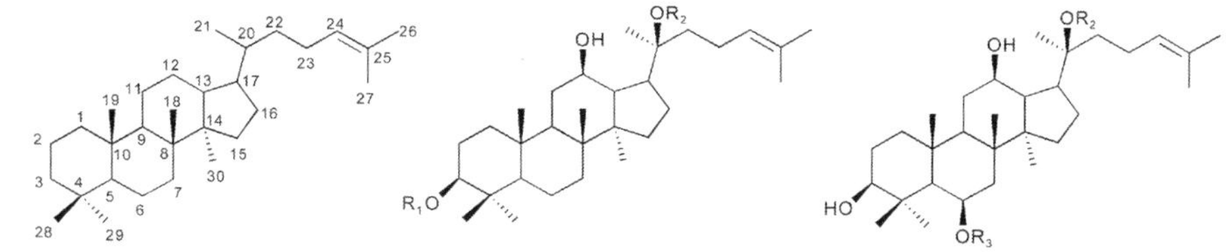

Damarane carbon skeleton | 20(S)-Protopanaxadiol carbon skeleton | 20(S)-Protopanaxatriol carbon skeleton

Ginsenoside	R_1	R_2	R_2	R_3
Rb1	Glc(1→2)Glc-	Glc(1→6)Glc-		
Rb2	Glc(1→2)Glc-	Ara(p) (1→6)Glc-		
Rc	Glc(1→2)Glc-	Ara(f) (1→6)Glc-		
Rd	Glc(1→2)Glc-	Glc-		
Gypenoside XVII	Glc-	Glc(1→6)Glc-		
Re			Glc-	Glc(1→2)Glc-
Rg1			Glc-	Glc-

Fig. 2.2 Structures of common ginsenosides. The core damarane structure common to ginsenosides is shown along with the carbon skeletons of the 20(*S*)-protopanaxadiol and 20(*S*)-protopanaxatriol ginsenosides. The sugar decorations of the most common ginsenosides are indicated in the table below the structures. Legend: Ara(*f*), arabinose (furanose form); Ara(*p*), arabinose (pyranose form); Glc, glucose

are based on a 3,12,20-trihydroxylated-20(*S*)-protopanaxadiol aglycone, with sugars attached to the hydroxyl groups at positions 3 and 20. Ginsenosides Rg_1 and Re, on the other hand, are based on a 3,6,12,20-tetrahydroxylated-20(*S*)-protopanaxatriol aglycone, with sugars attached to the hydroxyl groups at positions 6 and 20.

Biosynthetically, ginsenosides and sterols share a common intermediate precursor, oxidosqualene. Whereas cycloartenol synthase catalyzes the first committed step in sterol production, the enzyme dammarenediol synthase catalyzes the alternative folding and cyclization of 2,3-oxidosqualene into dammarenediol, the precursor of the dammarane skeleton of the ginsenosides [25, 26]. Unlike other classes of terpenoids, such as the chloroplast-derived monoterpenoids and diterpenoids, triterpenoids are a product of the cytosolic mevalonate pathway [27]. Newer results from the overexpression of squalene synthase [28], as well as interference on dammarenediol synthase [29] and cycloartenol synthase [30], have improved our understanding of the regulation of ginsenoside biosynthesis and demonstrated the feasibility of increasing ginsenoside content in planta using biotechnological approaches.

2.3.2 Distribution of Ginsenosides Within the Plant

American ginseng (*P. quinquefolius* L.) is a native North American member of the Araliaceae, a family whose more than 800 species are found mostly in the tropics

[31]. *P. quinquefolius* is a perennial understory herb that is associated with deciduous forests [32, 33] and ranges from Ontario and Quebec, south to northern Florida, and west to Minnesota [22]. The aboveground tissues senesce at the end of each growing season and estimates of the maximum age of this plant are from 23 to 30 years [33] to greater than 50 years [34]. However, occurrences of wild populations of American ginseng are becoming increasingly rare, and virtually all American ginseng products currently on the market are derived from cultivated sources. American ginseng has a long history of cultivation in North America. For example, Proctor and Bailey [35] estimate that this plant has been commercially grown in Canada since the late nineteenth century under artificial shade or in woodlands. The main commercial product from both cultivated and wild ginseng alike is the root system, which is usually harvested after 3–5 years and is sought for its medicinal properties.

Ginsenosides have been demonstrated to occur throughout the plant including within the main roots, secondary roots, stem, leaves, flowers, and fruit [36, 37]. Although much of the research and medicinal uses have traditionally focused on ginseng roots, the foliage also contains high amounts of ginsenosides, sometimes at concentrations higher than those found in the roots [36, 37]. In American ginseng, the six main ginsenosides Rg_1, Re, Rb_1, Rb_2, Rc, and Rd represent between 3 and 5% of the dry weight of the roots. With analyses that include a more complete profile of ginsenosides, this value can reach over 5% [37]. The distribution of the ginsenosides in different plant organs is not uniform. For example, the highest concentrations of ginsenoside Rb_1 and Rd are found in the roots and leaves, respectively, whereas ginsenoside Re is found in relatively equal amounts throughout the plant [36–38].

2.3.3 Fungitoxicity of Ginsenosides

Even though American ginseng accumulates large amounts of structurally different saponins, this plant is attacked in commercial gardens by a diverse array of fungi. These include the foliar pathogens *Alternaria panax*, *Alternaria alternata*, and *Botrytis cinerea* and the root pathogens *Cylindrocarpon destructans*, *Rhizoctonia solani*, *Phytophthora cactorum*, *P. irregulare*, *P. ultimum*, and several species of *Fusarium*, including *Fusarium oxysporum* and *Fusarium solani* [39–43]. Although the pathogens in the family Pythiaceae (i.e., *Phytophthora cactorum*, *P. irregulare*, and *P. ultimum*) are superficially similar to and share the same nutritional mode as the other pathogens, these organisms are not true fungi. The Oomycota, a phylum that contains water molds and phytopathogens that cause downy mildews, is more related to golden algae, brown algae, and diatoms and belongs collectively with these organisms in the Stramenopila. Most of the other ginseng pathogens, as well as *Trichoderma* spp. that may act as agonists of the pathogens in the soil, are true fungi belonging to the ascomycetes.

The fungitoxic activity of ginsenosides on many of these phytopathogens and agonists have recently been completed [44, 45] and represent the most comprehensive research in this area to date. Ginsenosides were characterized as mildly fungitoxic, mainly based on a comparison to aescin, a mixture of saponins from horse chestnut. A clear selectivity in fungitoxicity was observed when ginsenosides were tested against true fungi in vitro. The growth of *Trichoderma* spp., as well as the leaf pathogen *A. panax*, was inhibited to a greater degree than the growth of the root pathogens *Fusarium* spp. and *C. destructans* [44]. For example, the fungus that exhibited the greatest growth inhibition was *Trichoderma harzianum* (over 26% less growth than control), whereas the fungus that exhibited the least growth inhibition was *C. destructans* (over 7% more growth than control) at concentrations of 1 mg/mL crude ginsenoside fraction [44]. In addition, similar results were obtained when the tests were repeated under different temperature and nutrient regimes (R.W. Nicol, PhD Thesis, The University of Western Ontario, 2003). Unexpected growth stimulation was observed when the oomycotan pathogens were exposed to ginsenosides. The colony weights of both *P. cactorum* and *P. irregulare* grown with 5 mg/mL crude ginsenoside fraction were found to be significantly higher than controls [45]. Thus, both the fungal and oomycete root pathogens of American ginseng exhibited greater resistance to the chemical defenses of this plant. Moreover, the growth of *P. irregulare* was clearly stimulated by ginsenosides in a concentration-dependent manner [45].

2.3.4 Ginsenosides in the Rhizosphere

Although plant roots are known to influence the composition of soil microbes [46, 47], the exact mechanisms remain unknown. Several groups [45, 48, 49] have since suggested that secondary metabolites exuded from plant tissue can influence the composition of rhizosphere microbes. For example, our initial investigations into the exudation of ginsenosides from American ginseng revealed that these secondary compounds are present in ginseng garden rhizosphere soil as well as in root exudates collected from plants grown in a greenhouse. LC-MS and HPLC analyses of rhizosphere soil demonstrated that ginsenosides were present in this material at a concentration of 0.06% [45]. Moreover, when this low and ecologically relevant concentration of ginsenosides was used in bioassays, the growth of *P. irregulare* was found to be significantly greater than control [45].

2.4 Some Pathogens Can Degrade/Detoxify Saponins

2.4.1 Overview of "Saponinases" from Plant Pathogens

"Saponinases" are pathogen-produced glycosidases able to hydrolyze plant saponins, often leading to their detoxification/degradation. Numerous pathogenic fungi produce saponinases and some of these enzymes have been shown to determine host specificity [50, 51]. The most studied is the cereal-infecting fungus *G. graminis* var. *avenae*. While this species normally produces an extracellular "avenacinase," which detoxifies the oat triterpenoid saponin avenacin A-1 [50–52], mutants of the same species that do not produce the enzyme were unable to infect the saponin-containing host [50].

Septoria avenae and *Stagonospora avenae* are other examples of pathogens that are able to detoxify oat saponins. However, these two pathogens infect the leaves, and unlike *G. graminis* (which encounters avenacins when infecting the roots), they encounter a different family of antifungal compounds: the 26-desglucoavenacosides [20]. Both species have been shown to metabolize these saponins via extracellular glycosidases; while *Septoria avenae* secretes an avenacosidase with dual activity (α-L-rhamnosidase and β-D-glucosidase) [53], *Stagonospora avenae* secretes an α-rhamnosidase and two β-glucosidases [54]. Even though *Stagonospora avenae* α-rhamnosidase and β-glucosidases are expressed during infection [54], and *Septoria* isolates unable to deglycosylate the 26-DGAs are nonpathogenic to oats [53], neither of the *Stagonospora* β-glucosidases or *Septoria* avenacosidase has been proven to be either required for resistance to 26-desglucoavenacosides or pathogenicity to oats [53, 54].

The tomato leaf spot fungus *Septoria lycopersici* produces an extracellular tomatinase capable of converting α-tomatine into β2-tomatine, which is much less toxic to fungi [55–58]. Targeted mutation of the tomatinase gene in *S. lycopersici* resulted in the loss of the pathogen's ability to degrade α-tomatine and an increase in its sensitivity to the saponin [56, 58]. The ability of *S. lycopersici* to infect *Nicotiana benthamiana* (a solanaceous species that produces saponins) was shown to require the presence of tomatinase [59]. Further, the hydrolysis products were found to suppress the induced defense responses in plants by interfering with signal transduction processes associated with disease resistance [59]. The expression of tomatinase in *Nectria haematococca*, which resulted in its ability to detoxify α-tomatine and infect green tomato fruits [58, 60], also confirmed the involvement of this saponinase in pathogenicity.

Although the avenacinase of *G. graminis* var. *avenae* and the tomatinase of *S. lycopersici* are specific for their respective host saponins, both enzymes hydrolyze terminal β-1,2-linked D-glucose molecules and have similar mechanisms of action [55]. Sequence analysis of tomatinase has established that this enzyme belongs to the same family of β-glucosyl hydrolases as *G. graminis* avenacinase [55]. Other tomato-infecting fungi such as *Alternaria solani* [61], *Verticillium albo-atrum* [62], *F. oxysporum* f. p. *lycopersici* [63–65], *F. solani* [66], and *B. cinerea* [67] have also been reported to produce saponinases that deglycosylate the glycoalkaloid α-tomatine. Moreover, the causal agent of gray mold disease in many crops and vegetables, *B. cinerea*, is believed to possess at least three distinct saponin-specific glycosidases (a xylosidase and two different glucosidases), enabling it to detoxify avenacin, avenacosides, and digitonin [67, 68].

Although tomatinases from different fungal sources share the same substrate, they cleave different sugar moieties from α-tomatine. For example, tomatinases of *S. lycopersici* [55] and *V. albo-atrum* [62] remove the terminal β-1,2-linked glucose molecule, while that from *B. cinerea* removes the terminal β-1,3-linked xylose [67]. On the other hand, *F. oxysporum* sp. *lycopersici* [63, 64], *F. solani* [66], and

A. solani [61] remove the β-1-linked galactose and release the aglycone tomatidine along with the tetrasaccharide β-lycotetraose (*F. oxysporum* and *F. solani*), or the four individual monosaccharides (*A. solani*).

Gibberella pulicaris, a fungus that causes potato dry rot, is also able to detoxify saponins, particularly α-chaconine and α-solanine from potato, by removing the α-1, 2-L-rhamnose moieties of these molecules [69]. A chaconinase from this fungus has been purified, and although it was shown to be induced by its substrate, the enzyme's importance to virulence was not yet determined [70]. *G. pulicaris* was also shown to deglycosylate α-tomatine, by removing the β-lycotetraose as described for *F. oxysporum* f. sp. *lycopersici* [20, 71].

More recently, saponinases that are able to hydrolyze soybean saponins have been purified from the soybean pathogen *Neocosmospora vasinfecta* [72] and the saprotrophs *Aspergillus oryzae* and *Eupenicillium brefeldianum* [73]. While the saponinase of *N. vasinfecta* is considered to be required for detoxification of soybean saponins, the saponin hydrolases from *A. oryzae* and *E. brefeldianum* do not induce defense responses in plant cells [73]. Consequently, while saponin deglycosylation may be a component of the infection process of some organisms, it may not be sufficient in and of itself for pathogenicity. Even though the *N. vasinfecta*, *A. oryzae*, and *E. brefeldianum* saponin hydrolases belong to the same glycosidase family, they possess different specific activities, reflecting the organism's adaptation to different environments [73].

The majority of the fungal saponin-detoxifying glycosidases that have been characterized to date belong to family 3 of glycosyl hydrolases, along with cellobiose-degrading enzymes from yeasts and *Trichoderma* species [74]. Still other tomatinases, such as those secreted by *F. oxysporum* f. sp. *lycopersici* and *G. pulicaris*, belong to family 10 of glycosyl hydrolases, among many fungal xylanases [20, 60].

Unlike fungal saponinases, little is known about hydrolytic enzymes secreted by phytopathogenic oomycetes or their role in pathogenicity [75, 76]. Although innate resistance is believed to be the main mechanism by which oomycetes avoid the toxicity of saponins (see above), some oomycetes have been found to produce saponin hydrolases [77], as well as other glycosyl hydrolases [76].

2.4.2 The Metabolism of Ginsenosides by Pathogens of Ginseng

As noted above, ginsenosides are mildly fungitoxic against leaf pathogens and non-pathogenic fungi, but appear to promote the growth of more virulent root pathogens such as *P. irregulare* and *C. destructans*. Since the fungitoxic properties of some saponins are reduced through enzymatic degradation by pathogen-produced glycosidases, we explored the possibility that pathogens of ginseng produce glycosidases that degrade ginsenosides (R.W. Nicol, PhD Thesis, The University of Western Ontario, 2003) [77]. Accordingly, we cultured pathogenic (*P. irregulare*, *C. destructans*, *P. cactorum*) and nonpathogenic (*Trichoderma hamatum*) organisms in the presence of ginsenosides in vitro, and after several days of growth, profiled the ginsenosides that could be recovered from the spent medium. While minor alterations in the profile of ginsenosides recovered from the nonpathogenic fungus *T. hamatum* (e.g., decrease in Rb_1 and gypenoside XVII and an increase in Rd and F2) were observed, the overall profile was nearly identical to that of the profile of ginsenosides added to the medium at the start of the incubation (Fig. 2.3a,b). By contrast, the profile of ginsenosides recovered from cultures of *P. irregulare* (Fig. 2.3c–e), *C. destructans*, and *P. cactorum* (R.W. Nicol, PhD Thesis, The University of Western Ontario, 2003) were both quantitatively and qualitatively different from the initial ginsenoside mixture added to the culture medium. These alterations in ginsenoside profiles suggested either the degradation of ginsenosides or the presence of ginsenoside-hydrolyzing enzymes in the culture medium. Interestingly, the ability of different *P. irregulare* isolates to alter ginsenoside profiles was different (Fig. 2.3c–e). Two isolates (BR 598 and BR 1068), originally obtained from cultivated ginseng plants, nearly completely degraded all the 20(*S*)-protopanaxadiol ginsenosides (Fig. 2.3d, e), while a third isolate (BR 486; obtained from bean) showed limited activity (Fig. 2.3c).

2.4.3 *P. irregulare* Specifically Deglycosylates 20(S)-Protopanaxadiol Ginsenosides into Ginsenoside F2

One striking feature of the ginsenoside profiles obtained from the spent medium of *P. irregulare* cultures incubated with a crude mixture of ginsenosides is the apparent selectivity toward 20(*S*)-protopanaxadiol ginsenosides (Fig. 2.3). That is, in all profiles, the amount of 20(*S*)-protopanaxatriols (Rg_1, Re) remained unchanged, whereas significant quantitative changes were noted for all 20(*S*)-protopanaxadiol ginsenosides. While the amounts of most 20(*S*)-protopanaxadiol ginsenosides were depleted, one compound became enhanced in the profiles (Fig. 2.3d, e). The identity of this compound was confirmed as the 3,20-*bis*(*O*-β-D-glucose)-substituted 20(*S*)-protopanaxadiol ginsenoside F2, by a combination of LC-MS, ^{1}H-, and ^{13}C-NMR as well as negative ion FAB-MS [77]. Examination of the structures of the major 20(*S*)-protopanaxadiol ginsenosides reveals that F2 represents a structural core that could arise via simple deglycosylation of the 20(*S*)-protopanaxadiol ginsenosides Rb_1, Rb_2, Rc, and Rd (Fig. 2.4). Based on the existence of saponinases in other organisms that degrade saponins (see above), we concluded that *P. irregulare* also produces saponinases, which include specific ginsenoside-degrading ones; we termed these ginsenosidases.

2.4.4 *P. irregulare* Ginsenosidases Are Induced by Exposure to Ginsenosides In Vitro

Ginsenosidases of *P. irregulare* appear to be induced by their substrate in vitro. For example, when *P. irregulare* is incubated in the presence or absence of

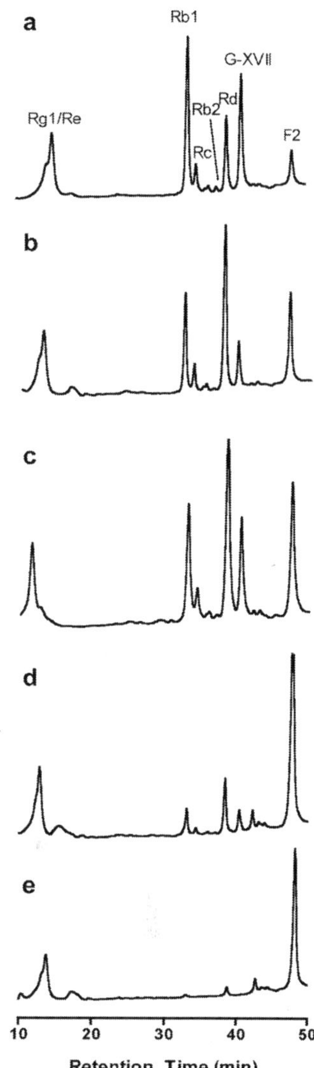

Fig. 2.3 HPLC analysis of ginsenosides recovered from the spent medium of *Trichoderma hamatum* and *Pythium irregulare* isolates. A ginsenoside mixture extracted from 3-year-old ginseng roots was added to the culture medium of both *T. hamatum* and *Pythium irregulare*, recovered after several days of incubation and analyzed by HPLC. The profile of ginsenosides added to the culture medium (**a**) included ginsenosides Rg_1, Re, Rb_1, Rb_2, Rc, Rd, and F2, as well as gypenoside XVII (G-XVII). (**b–e**) Profiles of ginsenosides recovered from *T. hamatum* isolate 3–323 (**b**), and *Pythium irregulare* isolates BR 486 (**c**), BR 598 (**d**), and BR 1068 (**e**). Adapted from M.A. Neculai, MSc Thesis, The University of Western Ontario, 2008

ginsenosides and screened for ginsenosidase activity the cultures pre-exposed to the ginsenosides show higher ginsenosidase activity (L.F. Yousef, MSc Thesis, The University of Western Ontario, 2005) [78]. Interestingly, both pre-exposed cultures and those that had not been exposed to ginsenosides (termed naïve) secreted proteins with *glycosidase* activity (measured with the artificial substrate *p*-nitrophenyl-β-D-glucoside, pNPG); however, the former secreted two to three times more proteins and ten times more glycosidase activity than naïve cultures [78]. Although in this case the two cultures had very similar glycosidase profiles, only glycosidases from cultures pre-exposed to ginsenosides readily deglycosylated the ginseng saponins

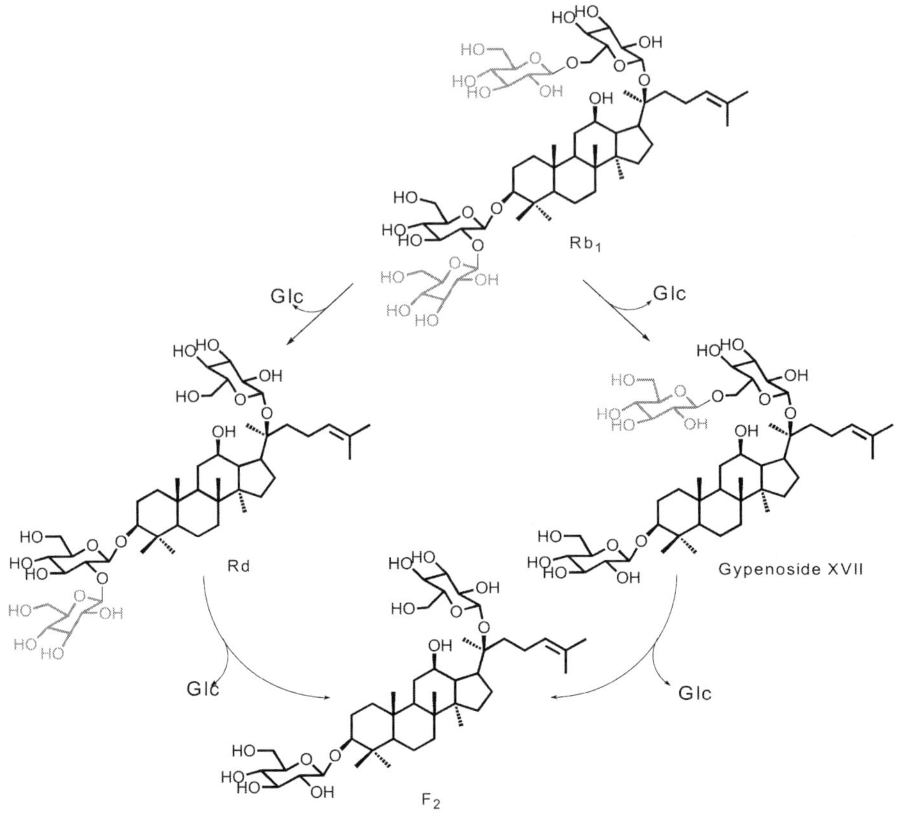

Fig. 2.4 Structural relationship between common 20(*S*)-protopanaxadiol ginsenosides and ginsenoside F2. The common core structure of ginsenoside F2 is highlighted in bold (*black*) within the structures of ginsenosides Rb₁, Rd, and gypenoside XVII. The sugar molecules that are cleaved off by hydrolysis by extracellular glycosidases are shaded in *gray*. In order to convert common 20(*S*)-protopanaxadiol ginsenosides into the common ginsenoside F2 product, the combined activity of both $\beta(1\rightarrow2)$ and $\beta(1\rightarrow6)$ glycosidases is required. Two additional 20(*S*)-protopanaxadiol ginsenosides, Rb₂ and Rc (data not shown), which contain an arabinose-glucose disaccharide at position C-20, could also feed into the production of ginsenoside F2; however, it is not clear at this time whether the same glycosidase that converts Rb₁ into Rd can also hydrolyze the arabinose $\beta(1\rightarrow6)$ glucose linkages of Rb₂ and Rc

[78]. In a manner similar to that reported for glycosidases of *G. pulicaris* [70] and *F. oxysporum* f. sp. *lycopersici* [64], we conclude that the secretion of ginsenosidases by *P. irregulare* is induced by the presence of the substrate. In other words, little or no saponinase activity was present in the filtrates of cultures pre-exposed to ginsenosides for 3–6 days, even though glycosidases were present. After that, ginsenosidases were found in relative abundance [78], suggesting an inductive process. Although these data may indicate involvement of ginsenosidases in the pathogenicity of *P. irregulare* toward American ginseng, further studies involving purified enzymes had to be undertaken.

2.4.5 Purification and Characterization of Ginsenosidases from P. irregulare

Three ginsenoside-deglycosylating enzymes were purified 13- to 25-fold from *P. irregulare* cultures grown in the presence of ginsenosides, by employing acetone precipitation (50% v/v), chromatofocusing between pH 7 and 4, gel filtration on Sephacryl S-200, and ion-exchange chromatography on Q Sepharose Fast Flow [78]. Chromatofocusing indicated the presence of more than one glycosidase in the sample or of different isoforms of the same glycosidase, similar to the tomatinase isolated from *F. oxysporum* f. sp. *lycopersici* [64], and predicted the isoelectric points of the enzymes to be 4.5–5.0. Gel filtration separated two distinct glycosidase activities, designated G1 and G2. Ion-exchange chromatography further purified the two glycosidase activities and electrophoretic analysis of the final ion-exchange fractions identified three enzymes: G1, which had eluted in the first peak with glycosidase activity after gel filtration (i.e., the G1 fractions), and G2/3, two enzymes similar in size, which had both eluted in the second peak with glycosidase activity from gel filtration (i.e., the G2 fractions) [78].

The molecular weights of *P. irregulare* ginsenosidases were estimated by gel filtration and SDS-PAGE. While gel filtration estimated 161 kDa for G1 and 46 kDa for G2/3, SDS-PAGE revealed a 78 kDa protein for G1, and two proteins of 61 and 57 kDa, respectively, for G2 and G3. It was concluded that G1 is a homodimer with two subunits of 78 kDa, and G2/3 are two monomeric enzymes [78]. Overall, the molecular weights and pI values of *P. irregulare* glycosidases were found to be similar to those of other characterized ginsenosidases such as the β-glucosidase secreted by *Panax ginseng* [79], the arabinopyranosidase and arabinofuranosidase of *Bifidobacterium breve* [80], and the β-glucosidase from *Fusobacterium* [81]. Moreover, these properties were also shared by some of the fungal saponinases (e.g., the tomatinase and avenacinase of *B. cinerea* [67, 68], the tomatinase of *F. oxysporum* sp. *lycopersici* [65]) previously characterized.

Although both purified glycosidase fractions (i.e., G1 and G2/3 obtained after ion-exchange chromatography) were found to deglycosylate ginsenosides, differences in the catalytic mechanisms were observed. While G1 showed β(1→6) hydrolytic activity, cleaving the terminal glucose from the disaccharide at C-20 of ginsenosides Rb$_1$ and G-XVII (to form Rd and F2, respectively), G2/3 showed β(1→2) hydrolytic activity, cleaving the terminal glucose from the disaccharide at C-3 of ginsenosides Rb$_1$ and Rd, with accumulation of G-XVII and F2, respectively [78]. These findings were confirmed by incubating partially purified enzyme preparations (i.e., culture filtrates subjected only to gel filtration) with a crude mixture of ginsenosides, but also with pure Rb$_1$ and Rd [78]. In comparison to G1 and G2/3 isolated from *P. irregulare*, the β-glucosidase of *Fusobacterium* showed both β-1→2 and β-1→6 activities able to deglycosylate ginsenoside Rb$_1$ to F2, and further transforming ginsenoside F2 into compound K (CK) [81]. Similar to G1, the β-glucosidase isolated from *P. bainier* was found to be able to convert ginsenoside Rb$_1$ to Rd, showing β-1→6 activity [82].

De novo sequence analysis of purified *P. irregulare* ginsenosidases and subsequent comparison of the generated peptide sequences against the NCBI nonredundant protein sequence database revealed a number of interesting findings. Firstly, G2 (61 kDa) and G3 (57 kDa), the enzymes that eluted in the same fractions after gel filtration, were found to be different enzymes, sharing only six peptide sequences out of the 150 and 130, respectively, generated by PEAKS [78]. Moreover, both G2 and G3 were found to share sequence similarities with β-glucosidases from a variety of organisms, but exhibited the highest homology to enzymes from plants and bacteria, all classified in the glycosyl hydrolase family-1 [78]. On the other hand, only 20 peptide sequences were generated for G1 (78 kDa) and none of them showed similarities with known glycosidases [78].

2.5 The Involvement of Ginsenosides and Ginsenosidases in the Ginseng-*P. irregulare* Pathosystem

P. irregulare glucosidases are induced in the presence of ginsenosides and these enzymes catalyze the conversion of 20(*S*)-protopanaxadiol ginsenosides to ginsenoside F2, a metabolite with monosaccharides (glucose) at C-3 and C-20 of the sapogenin [78]. It is still unclear how (or if) these reactions contribute to the growth stimulation of *P. irregulare* observed in vitro [44]; however, we are working under the assumption that enzymatic conversion of ginsenosides into a common metabolite F2 could increase pathogen growth in a manner similar to that observed in vitro for ergosterol (on *P. irregulare*) [77] or soybean sterols (*Phytophthora sojae*) [83]. In addition, glucosidases converting ginseng saponins to ginsenoside F2 may be a part of the pathogen's host recognition processes. Both of these scenarios, elaborated on below, would have important implications for the severity of disease in ginseng.

Generally, oomycetes are unable to synthesize sterols because of deficiencies at key points in the sterol pathway [84, 85] and they are therefore dependent upon exogenous sterols for growth and development. These organisms acquire sterols from the environment through the action of the sterol-carrying proteins, known as elicitins [86–88]. Ginsenosides are structurally and biosynthetically similar to sterols; therefore it is possible that the sterol-binding (elicitin-like) proteins of *Pythium* spp. [89, 90] transport ginsenoside metabolites into the hyphae of the pathogen. The action of elicitin-like proteins, coupled with specific glycosidase activity, and the recovery of ginsenoside F2 from the hyphae of *P. irregulare* [77] could account for the growth stimulation of *P. irregulare* observed in vitro [44]. The induction of G1 and G2/3 in cultures supplemented with ginsenosides, but not in naïve cultures, although the latter also produce glycosidases [78] supports the idea that the formation of ginsenosidases by *P. irregulare* may represent an important step in the recognition of host roots by this pathogen. This is further supported by the observation that *P. irregulare* isolates obtained from different plant sources appear to have different levels of ginsenosidase activity in vitro (see Section 2.3.2).

As a working *hypothesis*, the role of ginsenosides and ginsenosidases in the ginseng–*P. irregulare* pathosystem may be conceptualized as follows: ginseng plants produce ginsenosides, which, over time, are released into the rhizosphere. *P. irregulare* secretes broad specificity glycosidases into its surroundings, which may encounter ginsenosides. The degradation of ginsenosides and the formation of products like ginsenoside F2 may signal the presence of ginseng roots to *P. irregulare* and lead to the expression of more specific glycosidases (i.e., with either $1 \rightarrow 2$ or $1 \rightarrow 6$ hydrolytic activity) by the pathogen. Alternatively, the small amount of F2 found in the ginsenoside profile of some roots may be sufficient to trigger the same response. Regardless, the subsequent, specific degradation of ginsenosides in the rhizosphere leading to a gradient in F2 concentration may lead *P. irregulare* to host roots (Fig. 2.5).

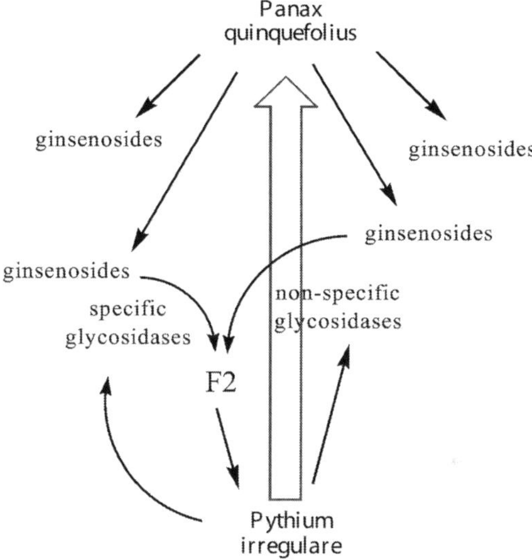

Fig. 2.5 Conceptual model for the role of ginsenosides in the interaction between ginseng and *Pythium irregulare*. Ginseng plants secrete/leach ginsenosides into the rhizosphere; similarly, *Pythium irregulare* secretes broad specificity glycosidases into its environment. When ginsenosides are encountered by glycosidases, at least one degradation product (F2) may be used by *Pythium irregulare* as a signal, resulting in the expression of more specific ginsenosidases by *Pythium irregulare*, which in turn generates more F2. Ultimately, *Pythium irregulare* moves toward the source of ginsenosides (*gray arrow*)

By contrast, given that *P. irregulare* tends to attack young seedlings (e.g., 1- or 2-year old plants), which may not have produced large quantities of ginsenosides, it may be that this pathogen does not use ginsenosides to find its host. Rather, it can be argued that *P. irregulare* uses mature plant secretions to locate a place where young seedlings will soon emerge; i.e., the presence of ginsenosides in the soil may lead to a build-up of inoculum that can infect subsequent plantings (whether natural or artificial). Such a hypothesis is supported by the "replant problem" associated with commercial ginseng production in which a new crop cannot be re-planted in the same gardens as previous ones [91]. Regardless, further experimental evidence is required to explicitly address this hypothesis.

2.6 Summary and Future Directions

The question posed in the title of this chapter remains to be answered. Are ginsenosides phytoanticipins? Or are they host recognition factors? Arguably, they are both. Since they are preformed antifungal compounds, they are by definition phytoanticipins. Their effectiveness is tempered by the fact that some organisms, but especially the oomycetes, are unaffected by their fungitoxicity, either via innate resistance or through active degradation. While such broad generalizations require further experimental validation, it has been shown that the ginseng pathogen *P. irregulare* can selectively deglycosylate the 20(*S*)-protopanaxadiol ginsenosides via extracellular glycosidases. We hypothesize that *P. irregulare* uses ginsenoside-specific glucosidases to help find its host and/or obtain nutrients/growth factors from its environment [78]. This hypothesis is supported by two unrelated observations: (1) different isolates of *P. irregulare* have been shown to have significant differences in pathogenicity between hosts [92] and (2) different isolates of *P. irregulare* have different levels of ginsenosidase activity in vitro (M.A. Neculai, MSc Thesis, The University of Western Ontario, 2008) (Fig. 2.3). If these observations are related, then the virulence of *P. irregulare* isolates on ginseng should correlate with ginsenosidase activity.

If the correlation between virulence and ginsenosidase acitivity exists, it may be feasible to generate avirulent strains of *P. irregulare* by knocking down these ginsenosidases in vitro. The corollary may also be true: if exposure to ginsenosides induces ginsenosidase activity (as suggested by in vitro data; [78]) and ginsenosidase activity is correlated with pathogenicity, then one might expect to increase a given isolate's pathogenicity toward ginseng (i.e., alter its host selectivity) by pre-exposing it to ginsenosides. This is supported by the observation that *P. irregulare* isolates are not necessarily more pathogenic on their host of origin but rather more selective in their pathogenicity toward different hosts [92]. This same trend of selective host pathogenicity based on genotypic relation has also been reported for *Pythium sylvaticum* and *P. ultimum* [93]. Establishing the selective host pathogenicity of *P. irregulare* will be vital in understanding the infection cycle of this oomycete and how exactly it is related to the metabolism of ginsenosides. This could inevitably lead to the establishment of better pathogen protection for ginseng growers whose crops are heavily affected by this pathogen.

References

1. Croteau RB et al (2000) Natural products (secondary metabolites). In: Buchanan, B, Gruissem, W, Jones, R (eds) Biochemistry and molecular biology of plants. American Society of Plant Physiologists, Rockville, MD, pp. 1250–1268

2. Harborne JB (1993) Introduction to ecological biochemistry, 4th edn. Academic Press, London

3. Dewick PM (1997) Medicinal natural products: a biosynthetic approach. Wiley, Chichester, UK

4. Grayer RJ, Harborne JB (1994) A survey of antifungal compounds from higher plants, 1982–1993. Phytochemistry 37:19

5. Sparg SG et al (2004) Biological activities and distribution of plant saponins. J Ethnopharm 94:219

6. Osbourn AE et al (2003) Dissecting plant secondary metabolism – constitutive chemical defences in cereals. New Phytol 159:101

7. Maizel JV et al (1964) Avenacin, an antimicrobial substance isolated from *Avena sativa*. I. Isolation and antimicrobial activity. Biochemistry 3:424

8. Maor R, Shirasu K (2005) The arms race continues: battle strategies between plants and fungal pathogens. Curr Opin Microbiol 8:399

9. Morant AV et al (2008) ß-Glucosidases as detonators of plant chemical defence. Plant Physiol 147:1072

10. Papadopoulou K et al (1999) Compromised disease resistance in saponin-deficient plants. Proc Natl Acad Sci USA 96:12923

11. Steel CC, Drysdale RB (1988) Electrolyte leakage from plant and fungal tissues and disruption of liposome membranes by α-tomatine. Phytochemistry 27:1025

12. Simons V et al (2006) Dual effects of plant steroidal alkaloids on *Saccharomyces cerevisiae*. Antimicrob Agents and Chemother 50:2732

13. Hostettmann K, Marston A (1995) Saponins. Cambridge University Press, Cambridge

14. Pareja-Jaime Y (2008) Tomatinase from *Fusarium oxysporum* f. sp *lycopersici* is required for full virulence on tomato plants. Mol Plant Microbe Interact 21:728

15. Cira LA et al (2008) Heterologous expression of *Fusarium oxysporum* tomatinase in *Saccharomyces cerevisiae* increases its resistance to saponins and improves ethanol production during the fermentation of *Agave tequilana* Weber var. azul and *Agave salmiana* must. Antonie Van Leeuwenhoek. Int J Gen Mol Microbiol 93:259

16. Keukens EAJ et al (1995) Molecular basis of glycoalkaloid induced membrane disruption. Biochim Biophys Acta 1240:216

17. Armah CN et al (1999) The membrane-permeabilizing effect of avenacin A-1 involves the reorganization of bilayer cholesterol. Biophys J 76:281

18. Stine KJ et al (2006) Interaction of the glycoalkaloid tomatine with DMPC and sterol monolayers studied by surface pressure measurements and brewster angle microscopy. J Physical Chem B 110:22220

19. Weete JD (1989) Structure and function of sterols in fungi. Adv Lipid Res 23:115

20. Morrissey JP, Osbourn AE (1999) Fungal resistance to plant antibiotics as a mechanism of pathogenesis. Microbiol Mol Biol Rev 63:708

21. Osbourn AE (1999) Antimicrobial phytoprotectants and fungal pathogens: a commentary. Fungal Genet Biol 26(3):163–168

22. Small E, Catling PM (1999) Canadian medicinal crops. NRC Research Press, Ottawa, ON

23. Fuzzati N et al (1999) Liquid chromatography-electrospray mass spectrometric identification of ginsenosides in *Panax ginseng* roots. J Chromatog A 854:69

24. Teng R et al (2002) Complete assignments of 1H and 13C NMR data for nine protopanatriol glycosides. Mag Res Chem 40:483

25. Pimpimon T et al (2006) Dammarenediol–II synthase, the first dedicated enzyme for ginsenoside biosynthesis, in *Panax ginseng*. FEBS Lett 580:5143

26. Liang Y, Zhao S (2008) Progress in understanding of ginsenoside biosynthesis. Plant Biol 10:415

27. Chappell J (2002) The genetics and molecular genetics of terpene and sterol origami. Curr Opin Plant Biol 5:151

28. Lee MH et al (2004) Enhanced triterpene and phytosterol biosynthesis in *Panax ginseng* overexpressing squalene synthase gene. Plant Cell Physiol 45:976
29. Han JY et al (2006) Expression and RNA interference-induced silencing of the dammarenediol synthase gene in *Panax ginseng*. Plant Cell 47:653
30. Liang Y et al (2009) Antisense suppression of cycloartenol synthase results in elevated ginsenoside levels in *Panax ginseng* hairy roots. Plant Mol Biol Rep 27:298
31. Lawrence GHM (1951) Taxonomy of vascular plants. Macmillan Press, New York, NY
32. Fountain MS (1986) Vegetation associated with natural populations of ginseng *Panax quinquefolium* in Arkansas USA. Castanea 51:42
33. Anderson RC et al (1993) The ecology and biology of *Panax quinquefolium* L. (Araliaceae) in Illinois. Am Midl Nat 129:357
34. Lewis WH, Zenger VE (1982) Population dynamics of the American ginseng *Panax quinquefolium* (Araliaceae). Am J Bot 69:1483
35. Proctor JTA, Bailey WG (1987) Ginseng, industry botany and culture. Hort Rev 9:187
36. Wills RBH et al (2002) Changes in ginsenosides in Australian-grown American ginseng plants (*Panax quinquefolium* L.). Aust J Exp Ag 42:1119
37. Qu CL et al (2009) Study on ginsenosides in different parts and ages of *Panax quinquefolius* L. Food Chem 115:340
38. Li TSC et al (1996) Ginsenosides in roots and leaves of American ginseng. J Agric Food Chem 44:717
39. Brammall RA (1994) Disappearing root rot. In: Howard, RJ, Garland, JA, Seaman, WL (eds) Diseases and Pests of Vegetable Crops in Canada. The Canadian Phytopathological Society and the Entomological Society of Canada, Ottawa, ON
40. Reeleder RD, Brammall RA (1995) Pathogenicity of *Pythium* species, *Cylindrocarpon destructans*, and *Rhizoctonia solani* to ginseng seedlings in Ontario. Can J Plant Pathol 16:311
41. Hill SN, Hausbeck MK (2008) Virulence and fungicide sensitivity of *Phytophthora cactorum* isolated from American ginseng gardens in Wisconsin and Michigan. Plant Dis 92:1183
42. Punja ZK (2000) Ginseng production in British Columbia and fungal diseases that affect yield and quality. Can J Plant Pathol 22:190
43. Punja ZK et al (2007) Diversity of *Fusarium* species associated with discolored ginseng roots in British Columbia. Can J Plant Pathol 29:340
44. Nicol RW et al (2002) Ginsenosides as host resistance factors in American ginseng (*Panax quinquefolius*). Can J Bot 80:557
45. Nicol RW et al (2003) Ginsenosides stimulate the growth of soilborne pathogens of American ginseng. Phytochemistry 64:257
46. Miethling R et al (2000) Variation of microbial rhizosphere communities in response to crop species, soil origin, and inoculation with *Sinorhizobium meliloti* L33. Microb Ecol 40:43
47. Grayston SJ et al (2001) Accounting for variability in soil microbial communities of temperate upland grassland ecosystems. Soil Biol Biochem 33:533
48. Broeckling CD et al (2008) Root exudates regulate soil fungal community composition and diversty. App Environ Microbiol 74:738
49. Saunders M, Kohn LM (2009) Evidence for alteration of fungal endophyte community assembly by host defence compounds. New Phytol 182:229
50. Bowyer P et al (1995) Host range of a plant pathogenic fungus determined by a saponin detoxifying enzyme. Science 267:371
51. Crombie WML et al (1986) Pathogenicity of the take-all fungus to oats; its relationship to the concentration and detoxification of the four avenacins. Phytochemistry 25:2075
52. Osbourn AE et al (1991) Partial characterization of avenacinase from *Gaeumannomyces graminis* var. *avenae*. Physiol Mol Plant Pathol 38:301
53. Wubben JP et al (1996) Detoxification of oat leaf saponins by *Septoria avenae*. Phytopathology 86:986
54. Morrissey JP et al (2000) *Stagonospora avenae* secretes multiple enzymes that hydrolyze oat leaf saponins. Mol Plant Microbe Interact 13:1041

55. Osbourn A et al (1995) Fungal pathogens of oat roots and tomato leaves employ closely related enzymes to detoxify different host plant saponins. Mol Plant Microbe Interact 6:971
56. Martin-Hernandez AM et al (2000) Effects of targeted replacement of the tomatinase gene on the interaction of *Septoria lycopersici* with tomato plants. Mol Plant Microbe Interact 12:1301
57. Sandrock RW, Van Etten HD (1998) Fungal sensitivity to and enzymatic degradation of the phytoanticipin α-lomatine. Phytopathology 88:137
58. Sandrock RW, Van Etten HD (2001) The relevance of tomatinases activity in pathogens of tomato: disruption of the β2-tomatinase gene in *Colletotrichum coccodes* and *Septoria lycopersici* and heterologous expression of the *Septoria lycopersici* β2-tomatinase in *Nectria haematococca*, a pathogen of tomato fruit. Physiol Mol Plant Pathol 58:159
59. Bouarab K et al (2002) A saponin-detoxifying enzyme mediates suppression of plant defences. Nature 418:889
60. Faure D (2002) The family-3 glycoside hydrolases: from housekeeping functions to host-microbe interactions. Appl Environ Microbiol 68:1485
61. Schönbeck F, Schlösser E (1976) Preformed substances as potential protectants. In: Heitefuss, R, Williams, PH (eds) Physiological plant pathology. Springer, Berlin, pp. 653–678
62. Pegg GF, Woodward S (1986) Synthesis and metabolism of α-tomatine in tomato isolines in relation to resistance to *Verticillium albo-atrum*. Mol Plant Microbe Interact 28:187
63. Ford JE et al (1977) The detoxification of α-tomatine by *Fusarium oxysporum* f.sp. *lycopersici*. Phytochemistry 16:544
64. Lairini K et al (1996) Purification and characterization of tomatinase from *Fusarium oxysporum* f. sp. *lycopersici*. Appl Environ Microbiol 62:1604
65. Roldán-Arjona TA et al (1999) Tomatinase from *Fusarium oxysporum* f.sp. *lycopersici* defines a new class of saponinases. Mol Plant Microbe Interact 12:852
66. Lairini K, Ruiz-Rubio M (1998) Detoxification of α-tomatine by *Fusarium solani*. Mycol Res 102:1375
67. Quidde T et al (1998) Detoxification of α-tomatine by *Botrytis cinerea*. Physiol Mol Plant Pathol 52:151
68. Quidde T et al (1999) Evidence for three different specific saponins-detoxifying activities in *Botrytis cinerea* and cloning and functional analysis of a gene coding for a putative avenacinase. Eur J Plant Pathol 105:273
69. Weltring KM et al (1997) Metabolism of potato saponins α-solanine and α-chaconine by *Gibberella pulicaris*. Phytochemistry 46:1005
70. Becker P, Weltring KM (1998) Purification and characterization of a-chaconinase of *Gibberella pulicaris*. FEMS Microbiol Lett 167:197
71. Weltring KM et al (1998) Metabolism of the tomato saponin α-tomatine by *Gibberella pulicaris*. Phytochemistry 48:1321
72. Watanabe M et al (2004) A novel saponin hydrolase from *Necosmosporsa vasinfecta*. Appl Environ Microbiol 70:865
73. Watanabe M et al (2005) Cloning and characterization of saponin hydrolases from *Aspergillus oryzae* and *Eupenicillium brefeldianum*. Biosci Biotechnol Biochem 69(11):2178
74. Osbourn A (1996) Preformed antimicrobial compounds and plant defence against fungal attack. Plant Cell 8:1821
75. Govers F et al (1997) The potato late blight pathogen *Phytophthora infestans* and other pathogenic oomycota. In: Caroll, HE Tudzynski, P (eds) The mycota part B plant relationships. Springer, Berlin, pp. 17–36
76. Brunner F et al (2002) A ß-glucosidase/xylosidase from the phytopathogenic oomycete, *Phytophthora infestans*. Phytochemistry 59:689
77. Yousef LF, Bernards MA (2006) In vitro metabolism of ginsenosides by the ginseng root pathogen *Pythium irregulare*. Phytochemistry 67:1740
78. Neculai MA et al (2009) Partial purification and characterization of three ginsenoside-metabolizing b-glucosidases from *Pythium irregulare*. Phytochemistry 70:1948

79. Zhang C et al (2001) Purification and characterization of a ginsenoside-β-glucosidase from ginseng. Chem Pharm Bull 49:795
80. Shin HY et al (2003) Purification and characterization of a α-L-arabinopyranosidase and α-L-arabinofuranosidase from *Bifidobacterium breve* K-110, a human intestinal anaerobic bacterium metabolizing ginsenoside Rb$_2$ and Rc. Appl Environ Microbiol 69:7116
81. Park SY et al (2001) Purification and characterization of ginsenoside Rb$_1$-metabolizing β-glucosidase from *Fusobacterium* K-60, a human intestinal anaerobic bacterium. Biosci Biotechnol Biochem 65(5):1163
82. Yan Q et al (2008) Purification method improvement and characterization of a novel ginsenoside-hydrolyzing-glucosidase from *Paecilomyces bainier* sp. 229. Biosci Biotechnol Biochem 72:352
83. Marshall JA et al (2001) Soybean sterol composition and utilization by *Phytophthora sojae*. Phytochemistry 58:423
84. Wood SG, Gottlieb D (1978) Evidence from mycelial studies for differences in the sterol biosynthetic pathway of *Rhizoctonia solani* and *Phytophthora cinnamomi*. Biochem J 170:343
85. Nes WD et al (1986) A comparison of cycloartenol and lanosterol biosynthesis and metabolism by the Oomycetes. Experientia 42:556
86. Mikes V (1997) The fungal elicitor cryptogein is a sterol carrier protein. FEBS Lett 416:190
87. Ponchet M et al (1999) Are elicitins cryptograms in plant-Oomycete communications? Cell Mol Life Sci 56:1020
88. Yousef LF et al (2009) Stigmasterol and cholesterol regulate the expression of elicitin genes in *Phytophthora sojae*. J Chem Ecol 35:824
89. Huet JC et al (1995) The relationships between the toxicity and the primary and secondary structures of elicitin-like protein elicitors secreted by the phytopathogenic fungus *Pythium vexans*. Mol Plant Microbe Interact 8:302
90. Tyler BM (2002) Molecular basis of recognition between *Phytophthora* pathogens and their hosts. Annu Rev Phytopathol 40:137
91. Kernaghan G,et al T. (2007) Quantification of *Cylindrocarpon destructans* f. sp *panacis* in soils by real-time PCR. Plant Pathol 56:508
92. Harvey PJ et al (2001) Genetic and pathogenic variation among cereal, medic and sub-clover isolates of *Pythium irregulare*. Mycological Res 105:85
93. Sewell GWF (1981) Effects of *Pythium* species on the growth of apple and their possible role in replant disease. Annals Appl Biol 97:31
94. Harborne JB, Baxter H (1995) Phytochemical dictionary. Taylor & Francis, London

Chapter 3
Fractionation of Grape Seed Proanthocyanidins for Bioactivity Assessment

Vaishali Sharma, Chungfen Zhang, Giulio M. Pasinetti, and Richard A. Dixon

3.1 Introduction

Proanthocyanidins (PAs), also known as condensed tannins, are polymeric flavonoid compounds composed of flavan-3-ol subunits and are widespread throughout the plant kingdom. They play a role in plant defense by accumulating in many different organs and tissues where they provide protection against predation [1]. PAs are the major class of phenolics in grape and grape-derived products such as wine, juice, and dietary supplements. These compounds are located in the skin and seeds and are responsible for imparting astringency to grape beverages [2]. Grape seed extract (GSE) is a complex mixture of PA monomers, oligomers, and polymers consisting of (+)-catechin, (–)-epicatechin, and (–)-epicatechin gallate subunits linked by C4→C8 or C4→C6 interflavan bonds (Fig. 3.1). The chemical structures of grape seed flavan-3-ols dictate their physical and biological properties. For example, the sensory perception of astringency and bitterness in wine changes in response to the degree of polymerization (DP) and degree of galloylation (DG) of the flavan-3-ols [2].

Recent studies indicate that PAs are strong antioxidants, and these compounds have been ascribed a number of potential activities beneficial to health, including protection against cancers, cardiovascular disease, and Alzheimers's disease [3–5]. As many as 5.3 million Americans and 27 million people worldwide are living with Alzheimer's disease, which is characterized by a build-up of proteins (plaques and tangles) in the brain, leading to loss of connection between neurons. Recent studies suggest that moderate red wine consumption reduces the incidence of Alzheimer's disease [6]. Moreover, grape-derived polyphenolic compounds may inhibit deposition of the beta-amyloid protein in spaces between the nerve cells, a process also called Aβ oligomerization [7]. Studies at Mount Sinai School of Medicine showed that dietary grape seed extract significantly reduced Alzheimer's disease-type

R.A. Dixon (✉)
Plant Biology Division, Samuel Roberts Noble Foundation, Ardmore, OK 73401, USA
e-mail: radixon@noble.org

D.R. Gang (ed.), *The Biological Activity of Phytochemicals*, Recent Advances
in Phytochemistry 41, DOI 10.1007/978-1-4419-7299-6_3,
© Springer Science+Business Media, LLC 2011

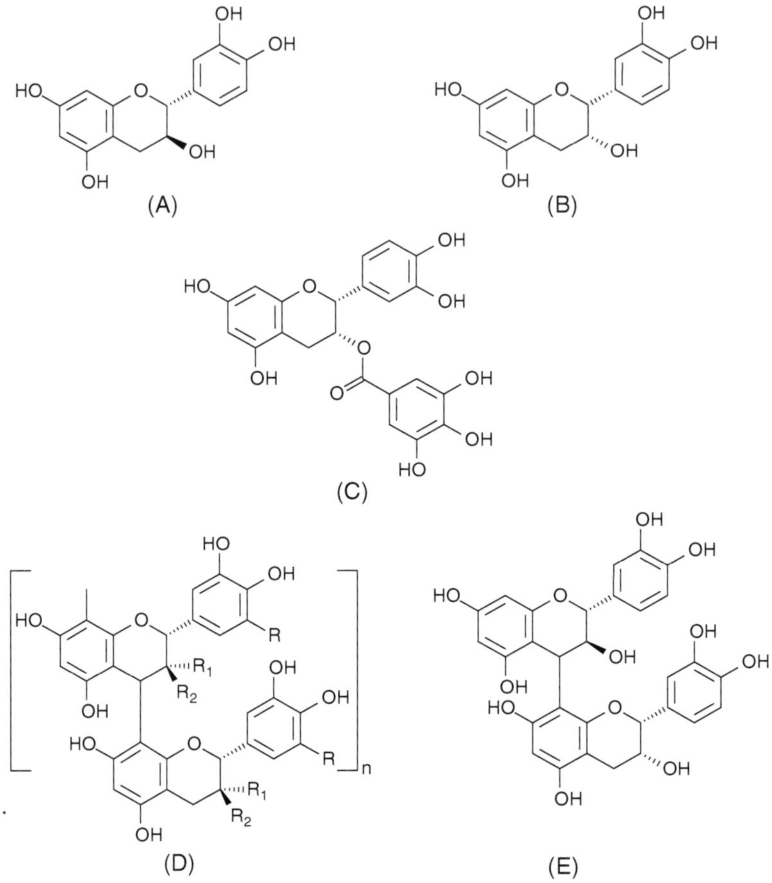

Fig. 3.1 Structures of (**a**) catechin, (**b**) epicatechin, (**c**) epicatechin gallate, (**d**) general structure of PAs, and (**e**) procyanidin dimer

cognitive deterioration in mice by prevention of amyloid formation in the brain [8]. However, the clinical relevance of these findings for GSE application in human diseases remains unknown, primarily because of our incomplete understanding of the metabolism and bioavailability of the polyphenolic components of GSE. Since GSE is a complex mixture of polyphenolics and other components, it is critical to purify the active fractions to the level of individual compounds or a small group of compounds in order to identify the bioactive compound(s) and determine whether there are synergies between molecular species.

The complete fractionation of PAs in plant material is somewhat difficult because of their structural complexity, in particular their different degrees of polymerization. Several methods ranging from fractionation on C18 Sep Pak cartridges, liquid–liquid extraction, gel permeation chromatography, and HPLC have been developed to separate oligomeric and polymeric PAs [8–11]. These methods generally allow a good separation of PA oligomers, but the resolution becomes poor with increasing degree of polymerization. In addition, irreversible adsorption can occur during chromatographic separations, particularly for polymeric species that interact strongly with adsorbent phases. On the other hand, a few methods based on liquid–liquid extraction [12] and precipitation [13] allow isolation of PAs from crude polyphenolic extracts, but the fractionation obtained is not sufficient to fully characterize a PA extract. Finally, direct analysis of PAs by ESI-MS without initial fractionation can identify oligomers and even polymers, but the size distribution remains undetermined [14].

The aim of this chapter is to highlight recent advances in fractionation techniques for grape seed PAs, using as examples methods we are using for large-scale fractionation into monomers, oligomers, and polymers for assessment of bioactivity and bioavailability in animal model studies of Alzheimer's disease. Our methods represent modifications/combinations of previously described approaches. We first discuss the available methods and modifications and then present a case study for the fractionation of MegaNatural-AZ GSE.

3.2 Methods for PA Extraction, Separation, and Analysis

3.2.1 Extraction of Grape Seed PAs

The total extractable phenolics in grapes are present at only about 10% in pulp, 60–70% in the seeds, and 28–30% in the skin. The most abundant phenolics isolated from grape seeds are catechins (catechin, epicatechin) and their polymers (PAs) [15, 16]. The solubility of PAs is governed by their chemical nature that may vary from simple to very highly polymerized substances. The usual methods for the isolation of PAs depend on their differential solubility in various solvents. Solvents such as acetone, methanol, ethanol, ethyl acetate, and their combinations, often with different proportions of water, have been used for the extraction of PAs from plant material [10, 17–19]. Other factors that play an important role in extraction include pH of the extraction medium, particle size of the biological material, temperature, and oxidation. Some researchers have reported the use of acidified solvents for the extraction of PAs from plant tissue [20, 21], while others have expressed concern that the presence of acids could induce degradation of PAs during extraction [22]. PAs from grape solids (seed, stem, skin) have been extracted at room temperature in the presence of nitrogen gas [23]. This reduces the degradation of PAs compared to extraction at high temperature in the presence of air (oxygen). Figure 3.2 shows the extraction efficiencies of grape seed PAs in different solvents [22].

3.2.2 Fractionation of PAs

Various methods have been reported for fractionation of PAs, among which solvent (liquid–liquid) extraction [10], column chromatography on Sephadex G25, LH20, or Toyopearl [8, 9], and solid phase extraction on C18 Sep-Pak cartridges

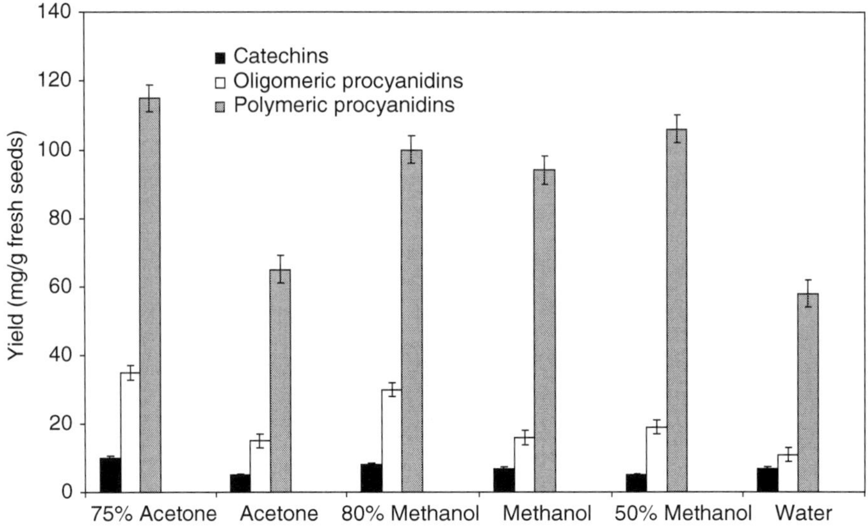

Fig. 3.2 Extraction efficiencies of grape seed proanthocyanidins in different solvents. Vertical bars represent standard deviation (adapted with permission from [23])

[8] are the most commonly used. Liquid–liquid extraction with ethyl acetate to isolate oligomeric PAs and with ether to isolate monomers has been used by several researchers [10]. However, the recovery is poor and only small amounts of PAs can be isolated in time-consuming separation steps. Therefore, the preparative isolation of PAs is most commonly achieved by employing Sephadex LH-20 and Toyopearl HW-40 column chromatography [8, 18, 20, 21, 24, 25]. The crude extract is applied to the column and non-tannin compounds are eluted with methanol followed by elution of PAs with aqueous acetone. According to Derdelinckx and Jerumanis [26], use of Toyopearl HW-40 allows PAs to be obtained in an advanced state of purity.

We have employed two different protocols for the chemical fractionation of GSE obtained from MegaNatural-AZ based on the amounts needed for bioactivity-based assays. Batches of GSE (50 g) were extracted in acetone/water (7:3) under N_2 with mechanical agitation for 12 h. The acetone was removed on a rotary evaporator and the aqueous phase was freeze-dried to yield 48 g of tannin crude extract (TCE). TCE was further fractionated following two different methods.

3.2.2.1 Solvent Precipitation

This fractionation of GSE follows the method of Saucier et al. [10]. TCE (2 g) was dissolved in Milli-Q water (10 g/L) and extracted three times with equal volumes of ethyl acetate. The organic phase (EA) was evaporated to dryness (35–40°C) and redissolved in 60 mL of water. The solution was then applied to an SPE Column (ENV1 18, 10 g, Supelco, Bellefonte, PA) and the monomers were eluted with diethyl ether. The oligomers were recovered by elution with 60 mL of methanol,

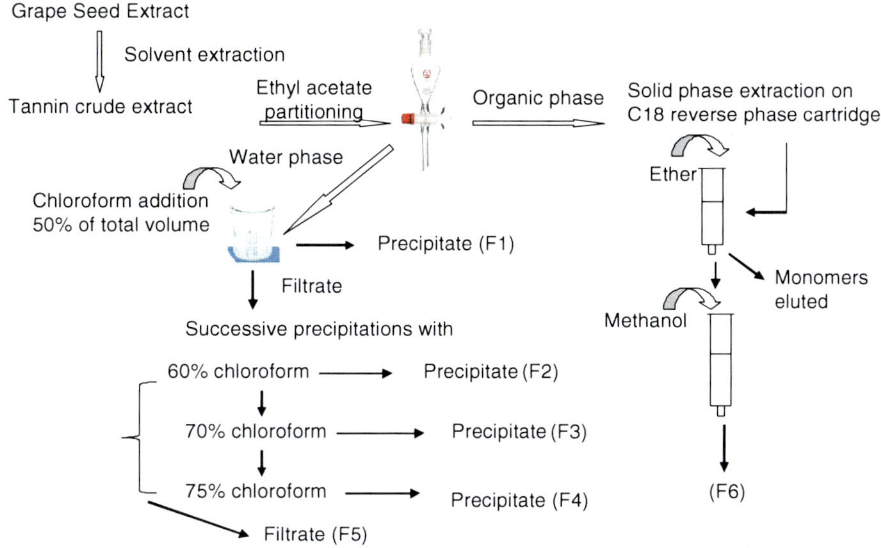

Fig. 3.3 Fractionation protocol for GSE based on successive solvent precipitations (adapted with permission from [10])

then evaporated to dryness, redissolved in a minimal volume of water, and freeze-dried to give fraction F6. The efficiency of monomer removal was checked by thin-layer chromatography on silica gel with ethyl acetate/formic acid (98:2, v/v) as mobile phase. The aqueous layer was evaporated to dryness and redissolved in 100 mL of methanol followed by addition of chloroform (100 mL). The precipitate that formed was collected by filtration and retained for the next steps. The methanolic extract was evaporated, redissolved in water, and freeze-dried to give fraction F1. This process was then repeated by adding successively 50 (60%), 83 (70%), and 66 (75%) mL of chloroform to the filtrate. This resulted in the F2, F3, and F4 precipitates and the final filtrate (F5). All these were evaporated, redissolved in water, and freeze-dried. Figure 3.3 shows the flow chart for fractionation of GSE by this procedure.

3.2.2.2 Fractionation on Toyopearl HW-40S

This protocol was developed for large-scale fractionation of grape seed PAs into specific size classes (monomer, oligomer, polymer). TCE (20 g) was dissolved in Milli-Q water (10 g/L) and extracted three times with equal volumes of ethyl acetate. The organic phase was evaporated to dryness (35–40°C) and redissolved in 60 mL of water. The solution was then applied to an SPE column (ENV1 18, 10 g, Supelco). The monomers were eluted with diethyl ether and evaporated to dryness, redissolved in a minimum volume of water, and freeze-dried to give the monomer fraction. The aqueous layer from the ethyl acetate extraction was evaporated to dryness and freeze-dried to give the aqueous extract. Fractionation was carried out by injecting

methanolic solutions containing 15 g of the aqueous extract material onto a 48 ×
4.8 cm glass column filled with Toyopearl HW-40S resin attached to an Isolera
flash purification system (Biotage, Charlottesville, VA). The compounds were eluted
with 50% methanol at a flow rate of 6 mL/min and monitored at 280 nm. Thirty
fractions each of 240 mL were collected, and the column was further eluted with
60% acetone and another 10 fractions of 240 mL were collected. Organic solvents
were evaporated at 35–40°C and the residue was freeze-dried. All fractions were
analyzed by normal phase (NP)-HPLC using the post-column derivatization method
described below. Figure 3.4 shows the flow chart for fractionation of GSE by this
method.

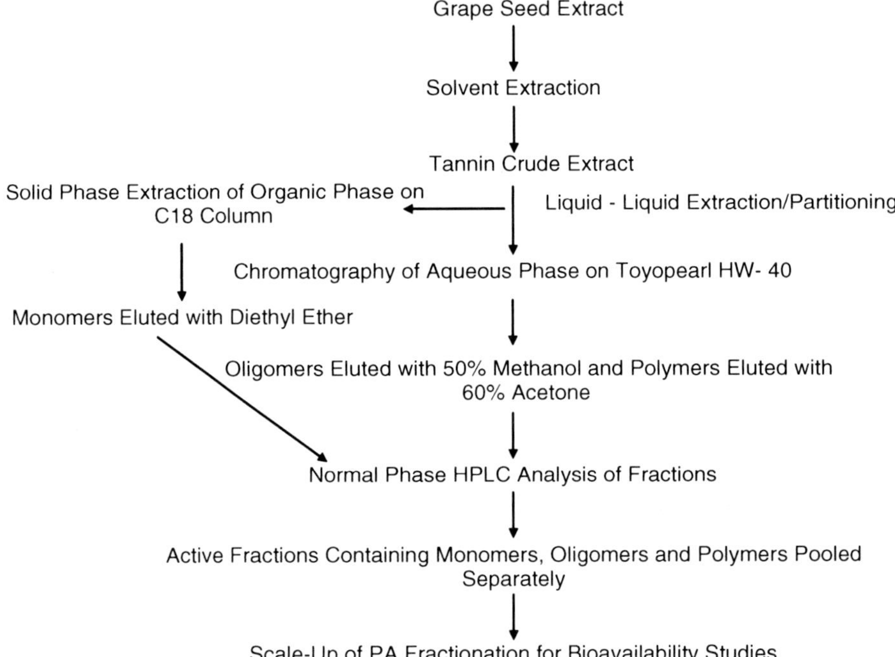

Fig. 3.4 Large-scale fractionation protocol for GSE on Toyopearl resin

3.2.3 Quantification of PAs

Methods used for the detection of PAs in crude or partially purified extracts can also
be adapted for post-column analysis after fractionation (see below). Direct quantitative analysis of PAs in crude grape phenolic extracts is often impossible due to
the complex sample matrix. Thus, fractionation or purification is often necessary
before analysis. The Folin-Ciocalteu and Prussian Blue assays are widely used for
the quantification of total polyphenols in plants [27, 28]. These methods are not specific for PAs due to the reaction of other phenolic compounds with these reagents.

Aldehydes are known to react with *m*-diphenols to form a colored carbonium ion in acid, and this reaction has been utilized for assay of flavanols because the A-rings of flavanols have *m*-diphenol functionality [29].

Vanillin, an aromatic aldehyde, has been used for PA assay [30], but vanillin protocols have, however, been reported to suffer from a lack of sensitivity, specificity, and reliability [31–33]. Another aldehyde, 4-dimethylaminocinnamaldehyde (DMACA), was reported to give a deep blue coloration after reaction with flavan 3-ols and is regarded as being specific for flavan 3-ols and their polymers (PAs) [34–36]. Figure 3.5 records this specificity by presenting standard curves for epicatechin and cyanidin-3-*O*-glucoside with DMACA reagent.

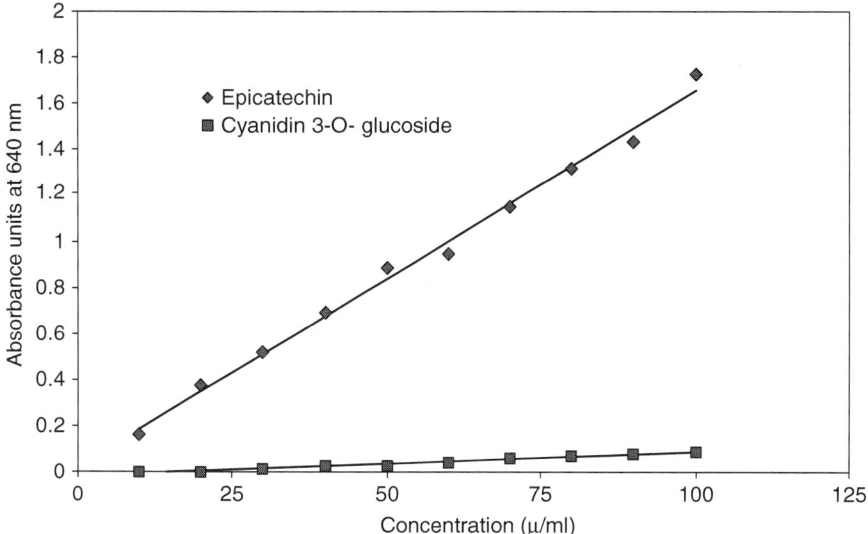

Fig. 3.5 Standard curves of epicatechin and cyanidin-3-*O*-glucoside with DMACA reagent

3.2.4 Determination of PA Complexity

High-performance liquid chromatography (HPLC) techniques are widely used for separation of phenolic compounds. Both reverse- and normal-phase HPLC methods have been used to separate and quantify PAs but have enjoyed only limited success. In reverse-phase HPLC, PAs smaller than trimers are well separated, while higher oligomers and polymers are co-eluted as a broad unresolved peak [8, 13, 37]. For our reverse-phase analyses, HPLC separation was achieved using a reverse phase, C18, 5 µm 4.6 × 250 mm column (J. T. Baker, http://www.mallbaker.com/). Samples were eluted with a water/acetonitrile gradient, 95:5 to 30:70 in 65 min, at a flow rate of 0.8 mL/min. The water was adjusted with acetic acid to a final concentration of 0.1%. All mass spectra were acquired using a Bruker Esquire LC-MS equipped with an electrospray ionization source in the positive mode.

Normal-phase HPLC has been used increasingly during the past few years for the separation of PAs from grape seed and cocoa [11]. PAs up to decamers were separated and resolved into equally spaced peaks in an order consistent with the degree of polymerization. However, for more complex polymers extracted from grape, the resolution becomes lower.

Several DMACA protocols have been developed for post-column detection of flavanols following HPLC analysis of beverages [31, 35, 36]. In our method, normal-phase HPLC analysis is performed using an HP 1100 system equipped with a diode array detector. Samples are analyzed using the post-column derivatization method described by Peel et al. [38]. Separations are done on a 250 × 4.6 mm Luna 5 μm silica column using methylene chloride and methanol in 4% acetic acid/water. Post-column derivatization is accomplished using a separate HPLC pump (Alltech model 426), which is used to deliver the DMACA reagent (1% DMACA in 1.5 M H_2SO_4 in methanol) to a mixing tee where effluent from the column and the reagent can combine and pass through an 8-m coil of 0.2 mm i.d. PEEK tubing prior to detection at 640 nm.

3.2.5 *Determination of the Mean Degree of Polymerization of PAs*

The degree of polymerization and nature of the flavan-3-ol subunits of PAs is usually determined by thiolysis or phloroglucinolysis. PAs under acidic conditions release terminal subunits as flavan-3-ol intermediates. These electrophilic intermediates can be trapped by a nucleophilic reagent to generate analyzable adducts. The two most commonly used nucleophilic reagents are phloroglucinol [25] and benzyl mercaptan [39]. The use of phloroglucinol as a trapping agent has two major advantages over benzyl mercaptan: (a) phloroglucinol is odorless and has no special handling requirements and (b) there is more selectivity in the formation of 3,4-*trans* adducts from 2,3-*trans* flavan-3-ol extension subunits [40].

For phloroglucinolysis, a solution of 0.1 N HCl in MeOH, containing 50 g/L phloroglucinol and 10 g/L ascorbic acid, is prepared. The PA of interest is reacted in this solution at 50°C for 20 min and then combined with 5 volumes of 40 mM aqueous sodium acetate to stop the reaction. After acid-catalyzed cleavage in the presence of phloroglucinol, the fraction is depolymerized and the terminal subunits released as flavan-3-ol monomers and the extension subunits released as phloroglucinol adducts of flavan-3-ol intermediates. These products are then separated and quantified by HPLC [25].

3.3 Case Studies for Fractionation of MegaNatural-AZ GSE

Dietary intervention with crude MegaNatural-AZ grape seed extract significantly reduced (by 30–50%) Alzheimer's disease-type cognitive deterioration in mice by prevention of amyloid formation in the brain [8]. However, it is essential to determine which components of the extract exhibited bioactivity in the transgenic mouse model of Alzheimer's disease and how this correlated with the bioavailability of the individual components. To this end, we have pursued the fractionation of GSE in quantities sufficient for analysis of the protective effects of its individual components against Alzheimer's disease in animal behavior models.

3.3.1 Solvent Precipitation

GSE was extracted with aqueous acetone to obtain a crude tannin extract (TCE). Normal-phase HPLC analysis of TCE (Fig. 3.6) showed that it is a mixture of monomers, oligomers, and polymers with a mean degree of polymerization of 4. This agrees with the results from previous studies [10, 41]. TCE was fractionated into six fractions by successive solvent extractions and precipitation (Fig. 3.3). Oligomers were extracted by ethyl acetate (fraction F6) and monomers were removed by solid phase extraction on C18 with diethyl ether according to the method of Saucier et al. [10]. However, monomer peaks still persist in the F6 fraction (Fig. 3.7f). The remaining polymers were dissolved in methanol and precipitated successively by adding increasing amounts of chloroform. Five fractions were recovered by vacuum filtration and analyzed on NP-HPLC with post-column derivatization with DMACA (Fig. 3.7a–e). The degree of polymerization of the fractions was obtained by phloroglucinolysis and is given in Table 3.1. The fractions have decreasing degree of polymerization with increasing fraction number, in accordance with the results of Saucier et al. [10]. This is a rapid method for the preparation of different molecular weight fractions of grape seed PAs, but there is still significant cross-contamination between the fractions. Moreover, individual fractions containing monomers, oligomers, and polymers were not separated by this method.

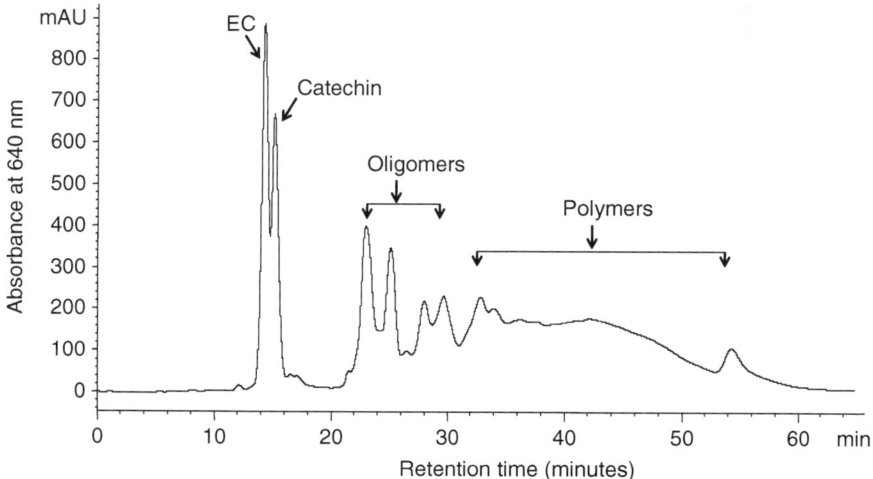

Fig. 3.6 Tannin crude extract from grape seed separated by normal-phase HPLC with post-column detection using DMACA

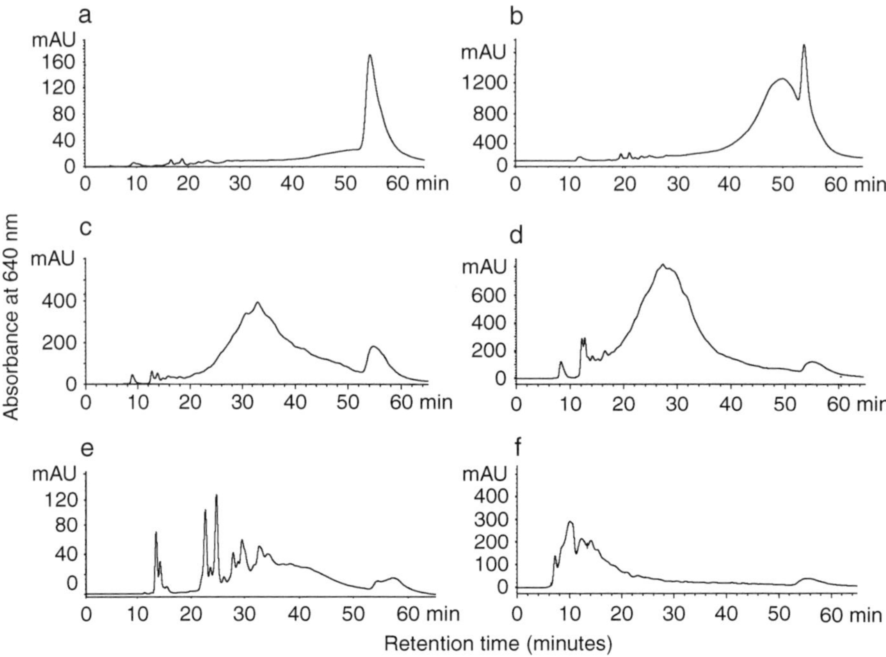

Fig. 3.7 Normal-phase-HPLC chromatograms of PA fractions generated by differential solvent extraction of crude GSE (protocol 1). (**a**) Fraction 1, (**b**) Fraction 2, (**c**) Fraction 3, (**d**) Fraction 4, (**e**) Fraction 5, and (**f**) Fraction 6. Compounds were detected with post-column derivatization using DMACA

Table 3.1 Mean degree of polymerization (mDP) of GSE fractions determined by phloroglucinolysis

	Mole %						
	EC-phloro adduct	Catechin-phloro adduct	ECG-phloro adduct	EGC-phloro adduct	Catechin	EC	mDP
TCE	52.6	5.0	13.4	2.4	12.7	13.5	4.0
Fraction 1[a]	53.0	8.1	15.0	5.9	10.2	11.6	4.5
Fraction 2[a]	52.0	8.9	14.0	1.1	10.9	13.0	4.1
Fraction 3[a]	51.4	8.9	14.8	1.6	11.2	13.4	4.0
Fraction 4[a]	53.2	7.4	11.9	1.3	11.2	16.0	3.6
Fraction 5[a]	47.1	6.9	13.8	1.2	13.4	18.6	3.1
Fraction 6[a]	40.7	6.5	14.8	1.4	18.0	19.8	2.6
Oligomer-rich[b]	39.0	5.5	4.87	1.5	20.0	29.0	2.0
Polymer-rich[b]	63.0	7.2	12.0	3.3	8.1	6.0	7.2

[a]GSE fractions from protocol 1.

[b]GSE fractions from protocol 2.

3.3.2 Fractionation on Toyopearl Resin

In order to obtain large amounts of monomers, oligomers, and polymers for bioactivity determination, we have employed another fractionation protocol combining different techniques of solvent extraction, partitioning, column chromatography on Toyopearl resin, and HPLC (Fig. 3.4). The purity of the fractions was confirmed by LC-MS. TCE was extracted repetitively with ethyl acetate and the organic layer containing monomers was further purified by solid phase extraction on a C18 cartridge with diethyl ether (Fig. 3.8a). Oligomers and polymers were fractionated from the aqueous layer of the TCE by column chromatography on Toyopearl resin. The three fractions were monomer-rich, oligomer-rich, and polymer-rich. NP-HPLC profiles of tannin crude extract, oligomer-rich fraction, polymer-rich fraction, and monomer-rich fraction are shown in Figs. 3.5 and 3.8 (a–d). The LC-MS analysis of the oligomer-rich fraction confirmed the presence of procyanidin B2 (Fig. 3.9). The oligomer-rich fraction has a mean degree of polymerization (mDP) of 2.0 and the polymer-rich fraction has an mDP of 7.2 as determined by phloroglucinolysis (Table 3.1, Fig. 3.10). There is little cross-contamination of the components in the monomer-rich, oligomer-rich, and polymer-rich fractions.

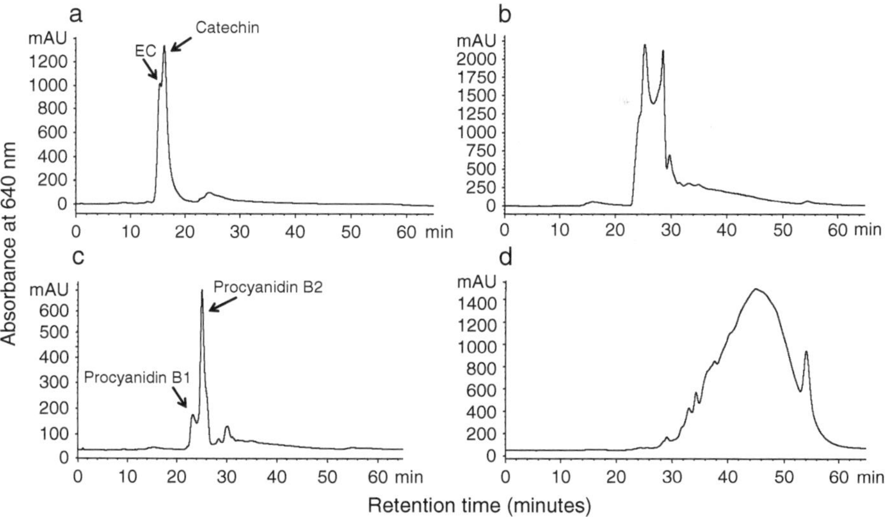

Fig. 3.8 Normal-phase-HPLC chromatograms of PA fractions generated by a combination of solvent extraction and column chromatography on Toyopearl resin (protocol 2). **a** monomer-rich, **b** oligomer-rich, **c** dimer-rich, and **d** polymer-rich fractions. Compounds were detected with post-column derivatization using DMACA

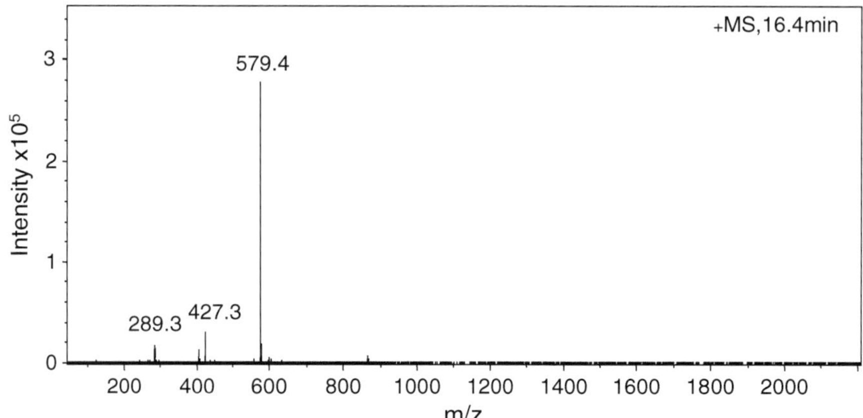

Fig. 3.9 ESI mass spectrum of the dimer-rich fraction from Fig. 3.8c showing the procyanidin B2 peak (this eluted at 16.4 min after reverse-phase HPLC separation)

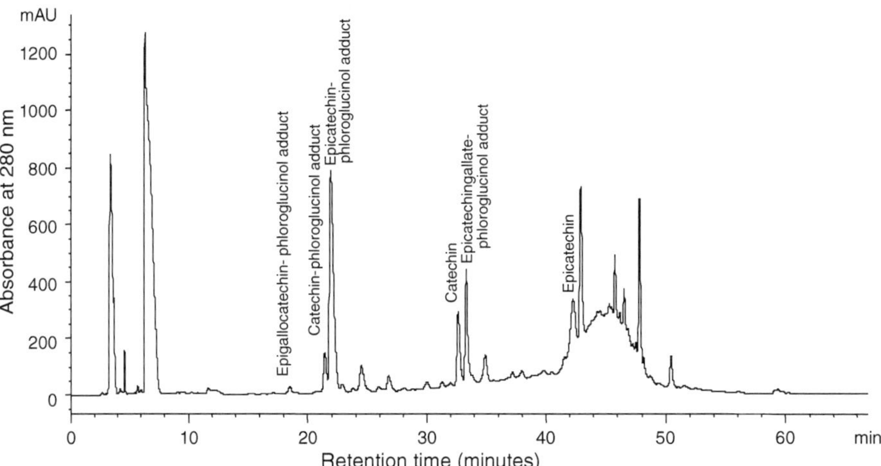

Fig. 3.10 Reverse-phase HPLC chromatogram of PA phloroglucinol cleavage products from the polymer-rich fraction of GSE

3.4 Conclusions

Crude GSE can be separated into distinct components from monomer-rich to oligomer- and polymer-rich fractions by solvent extraction followed by chromatography on Toyopearl resin. The pre-fractionation of the tannin crude extract by liquid–liquid extraction with ethyl acetate fractionates the monomers into the organic phase and improves the subsequent fractionation of oligomers and polymers on Toyopearl HW 40 resin. This fractionation technique is easy to operate and is reproducible. Furthermore, it can be achieved on a large-scale suitable for generation of sufficent material (5–10 g batches of specific size fractions) for bioactivity assays/animal feeding studies.

Acknowledgments We thank David Huhman for assistance with mass spectrometry analysis. This work was supported by the National Institutes of Health National Center for Complementary and Alternative Medicine (award # 1 PO1 AT004511) and the Samuel Roberts Noble Foundation.

References

1. Xie DY, Dixon RA (2005) Proanthocyanidin biosynthesis-still more questions than answers? Phytochemistry 66:2127–2144
2. Vidal SFL, Francis L, Guyot S et al (2003) The mouth-feel properties of grape and apple proanthocyanidins in a wine-like medium. J Sci Food Agric 83:564–573
3. Okello EJ, Savelev SU, Perry EK (2004) In vitro anti-beta-secretase and dual anti-cholinesterase activities of *Camellia sinensis* L. relevant to treatment of dementia. Phytother Res 18:624–627
4. Wang J, Ho L, Zhao W et al (2008) Grape-derived polyphenolics prevent Aβ oligomerization and attenuate cognitive deterioration in a mouse model of Alzheimer's disease. J Neurosci 28:6388–6392
5. Ricardo DS, Darmon N, Fernandez Y et al (1991) Oxygen free radical scavenger capacity in aqueous models of different procyanidins from grape seeds. J Agric Food Chem 39: 1549–1552
6. Ono K, Yoshiike Y, Takashima A et al (2003) Potent anti-amyloidogenic and fibril-destabilizing effects of polyphenols in vitro: implications for the prevention and therapeutics of Alzheimer's disease. J Neurochem 87:172–181
7. Porat Y, Abramowitz A, Gazit E (2006) Inhibition of amyloid fibril formation by polyphenols: structural similarity and aromatic interactions as a common inhibition mechanism. Chem Biol Drug Des 67:27–37
8. Sun B, Leandro C, Ricardo JM (1998) Separation of grape and wine proanthocyanidins according to their degree of polymerization. J Agric Food Chem 46:1390–1396
9. Ricardo DS, Rigaud JM, Cheynier V et al (1991) Procyanidin dimers and trimers from grape seeds. Phytochemistry 30:1259–1264
10. Saucier C, Mirabel M, Daviaud F et al (2001) Rapid fractionation of grape seed proanthocyanidins. J Agric Food Chem 49:5732–5735
11. Rigaud J, Escribano-Bailon MT, Prieur C et al (1993) Normal phase high performance liquid chromatographic separation of procyanidins from cacao beans and grape seeds. J Chromatogr 654:255–260
12. Kolodziej H (1985) Fractionation of condensed tannins by counter current chromatography. J Chromatogr 318:85–93
13. Giner-Chavez BI, Van Soest PJ, Robertson JB et al (1997) A method for isolating condensed tannins from crude plant extracts with trivalent ytterbium. J Sci Food Agric 74:359–368
14. Cheynier V, Doco T, Fulcrand H et al (1997) ESI-MS analysis of polyphenolic oligomers and polymers. Analysis 25:32–37
15. Silva JMR, Rigaud J, Cheynier V et al (1991) Procyanidin dimers and trimers from grape seeds. Phytochemistry 30:1259–1264
16. Prieur C, Rigaud J, Cheynier V et al (1994) Oligomeric and polymeric procyanidins from grape seeds. Phytochemistry 36:781–784
17. Sun B, Belchior P, Ricardo-da-Silva JM et al (1999) Isolation and purification of dimeric and trimeric procyanidins from grape seeds. J Chromatogr A 841:115–121
18. Gabetta B, Fuzzati N, Griffini A et al (2000) Characteriztion of proanthocyanidins from grape seeds. Fitoterapia 71:162–175

19. Yang Y, Chien M (2000) Characterization of grape procyanidins using high performance liquid chromatography/mass spectrometery and matrix assisted laser desorption/ionization time of flight mass spectrometry. J Agric Food Chem 48:3990–3996
20. Kennedy JA, Hayasaka Y, Vidal S et al (2001) Composition of grape skin proanthocyanidins at different stages of berry development. J Agric Food Chem 49:5348–5355
21. Labarbe B, Cheynier V, Brossaud F et al (1999) Quantitative fractionation of grape proanthocyanidins according to their degree of polymerization. J Agric Food Chem 47:2719–2723
22. Sun B, Spranger MI, Ricardo-da-Silva JM (1996) Extraction of grape seed procyanidins using different organic solvents. In: Vercauteren J, Cheze C, Dumon MC, Weber JF (eds) Proceedings of the 18th International Conference on Polyphenols. Groupe Polyphénols, Bordeaux (France)
23. Sun B, Spranger MI (2005) Review: quantitative extraction and analysis of grape and wine proanthocyanidins and stilbenes. Ciencia Tec Vitiv 20:59–89
24. Jing H, Deinzer ML (2007) Tandem mass spectrometry for sequencing proanthocyanidins. Anal Chem 79:1739–1748
25. Kennedy JA, Jones GP (2001) Analysis of proanthocyanidin cleavage products following acid-catalysis in the presence of excess phloroglucinol. J Agric Food Chem 49:1740–1746
26. Derdelinckx G, Jerumanis J (1984) Separation of malt hop proanthocyanidins on Fractogel TSK HW-40(S). J Chromatogr 285:231–244
27. Price ML, Butler LG (1977) Rapid visual estimation and spectrophotometric determination of tannin content of sorghum grain. J Agric Food Chem 25:1268–1273
28. Graham HD (1992) Stabilization of the Prussian blue color in the determination of polyphenols. J Agric Food Chem 40:801–805
29. Scalbert A (1992) Quantitative methods for the estimation of tannins in plant tissues. In: Hemingway RW, Laks PE (eds) Plant polyphenols: synthesis, properties, significance, vol 59. Plenum Press, New York, NY, pp 259–280
30. Broadhurst RB, Jones WT (1978) Analysis of condensed tannins using acidified vanillin. J Sci Food Agr 29:788–794
31. McMurrough I, McDowell J (1978) Chromatographic separation and automated analysis of flavanols. Anal Biochem 91:92–100
32. Butler LG, Price ML, Brotherton JE (1982) Vanillin assay for proanthocyanidins (condensed tannins): modification of the solvent for estimation of the degree of polymerization. J Agric Food Chem 30:1087–1089
33. Makkar HPS, Becker K (1993) Vanillin-HCl method for condensed tannins: effect of organic solvents used for extraction of tannins. J Chem Ecol 19:613–621
34. Thies M, Fischer R (1971) A new colour reaction for the microchemical detection and the quantitative determination of catechins. Mikrochim Acta 1:9–13
35. Treutter D (1989) Chemical reaction detection of catechins and proanthocyanidins with 4-dimethylaminocinnamaldehyde. J Chromatogr 467:185–193
36. Nagel CW, Glories Y (1991) Use of a modified dimethylaminocinnamaldehyde reagent for analysis of flavanols. Am J Enol Vitic 42:364–366
37. Guyot S, Doco T, Souquet JM et al (1997) Characterization of highly polymerized procyanidins in cider apple skin and pulp. Phytochemistry 44:351–357
38. Peel GJ, Dixon RA (2007) Detection and quantification of engineered proanthocyanidins in transgenic plants. Nat Prod Commun 2:1009–1014
39. Mathews S, Mila I, Scalbert A et al (1997) Method for estimation of proanthocyanidins based on their acid depolymerization in the presence of nucleophiles. J Agric Food Chem 45: 1195–1201
40. Hermingway RW (1989) Reactions at the interflavanoid bond of proanthocyanidins. In: Hermingway RW, Karchesy JJ (eds) Chemistry and significance of condensed tannins. Plenum Press, New York, NY
41. Kohler N, Wray V, Winterhalter P (2008) Preparative isolation of procyanidins from grape seed extracts by high-speed counter-current chromatography. J Chromatogr A 1177:114–125

Chapter 4
Coloring Soybeans with Anthocyanins?

Nikola Kovinich, John T. Arnason, Vincenzo De Luca, and Brian Miki

4.1 Anthocyanin Biosynthesis in Black Soybean for the Visual Identification of Transgenic Grains?

The commercial production of genetically engineered (GE) crops has risen to 250 million acres worldwide [1]. As the prevalence and diversity of engineered crops increase, contamination of non-modified grain with transgenics is becoming problematic [2]. Distinct colors for the visible identification and quantitation of transgenic materials may provide increased safety to the consumer and freedom to the producer. Black soybean, long considered a medicinal food and consumed only occasionally in Japan and Korea, is similar in composition to cultivated yellow soybean, except that it accumulates anthocyanin and proanthocyanidin (PA) pigments in the seed coat. Recent advances in the understanding of anthocyanin biosynthesis, composition, and flux in black soybeans may enable the engineering of distinct seed colors with low potential for unintended effects on the biosafety of food and feed. Since the first reported engineering of a "novel" flower color by the expression of a maize enzyme in petunia [3], numerous experiments have successfully modified flower color in various plant species [4]. However, successful engineering of target colors is likely uncommon, as exemplified by more than one decade of attempts to engineer a *true* blue rose [5, 6]. Thus, a better understanding of black color in plant tissues may enable the engineering of different, yet unpredictable colors from black soybeans.

N. Kovinich (✉)
Bioproducts and Bioprocesses, Research Branch, Agriculture and Agri-Food Canada, Ottawa, ON K1A 0C6, Canada; Department of Biology, Carleton University, Ottawa-Carleton Institute of Biology, Ottawa, ON K1S 5B6, Canada
e-mail: kovinichn@agr.gc.ca

D.R. Gang (ed.), *The Biological Activity of Phytochemicals*, Recent Advances in Phytochemistry 41, DOI 10.1007/978-1-4419-7299-6_4,
© Springer Science+Business Media, LLC 2011

4.2 Anthocyanins in Black Soybean

4.2.1 Anthocyanin Composition in the Seed Coat of Black Soybean

In black soybean seeds, anthocyanins accumulate within the seed coat palisade cells (the epidermal layer). Until 2009, only major anthocyanins within the black soybean seed coat had been reported. They were structurally simple, consisting of 3-*O*-glucosides of cyanidin, delphinidin, and petunidin in most varieties [7]. In 2009, Lee et al. [8] reported an in-depth analysis of anthocyanin composition from the seed coats of black soybean and identified six additional anthocyanins that included major amounts of pelargonidin-3-*O*-glucoside, catechin-cyanidin-3-*O*-glucoside, and minor amounts of cyanidin-3-*O*-galactoside, delphinidin-3-*O*-galactoside, cyanidin, and peonidin-3-*O*-glucoside. This suggests that anthocyanin biosynthesis in the seed coat is more complex than previously envisioned.

4.2.2 Anthocyanin Biosynthesis, Flux, and Accumulation
in Black Soybean

Despite the requirement for various regulatory, structural, and transport genes required for coordinate biosynthesis and accumulation of anthocyanins [9], only six genetic loci (*I*, *R*, *T*, *Wp*, *W1*, and *O*) have been reported to control seed coat color in soybean [10, 11]. These loci likely represent nonredundant regulators of anthocyanin biosynthesis. Chalcone synthase, the branch-point enzyme that commits phenylpropanoid flux to flavonoid/anthocyanin biosynthesis (Fig. 4.1), is encoded by a small gene family (*CHS1–CHS9*) [12]. *CHS7* and *CHS8* function in the seed coat [13, 14]. It is by silencing *CHS7* and *CHS8* specifically in the seed coat palisade via RNA interference [13, 14] that the *Inhibitor* (*I*) locus reduces flux into the flavonoid pathway causing the absence of visual pigments and a yellow seed appearance. There are four alleles of the *I* locus (*I*, *i^i^*, *i^k^*, and *i*), each of which results in different spatial distribution of pigments in the seed coat [10]. It is recessive allele *i* that does not silence *CHS7* and *CHS8*, allowing for pigment production throughout the seed coat palisade cells

---

Fig. 4.1 (continued) Anthocyanin biosynthetic pathway in the seed coat of *black* soybean. Only major anthocyanins are shown. Metabolites are represented in lowercase, enzymes in uppercase, locus alleles in shaded boxes. Major flux control points (large arrows). A circled plus indicates positive regulation, a vertical bar, inhibition, and a question mark, an unknown mechanism. AOMT, anthocyanin-*O*-methyltransferase; CHI, chalcone isomerase; CHS7, CHS8, chalcone synthases; DFR, dihydroflavonol 4-reductase; F3H, flavanone 3-hydroxylase; F3′H, flavonoid 3′-hydroxylase; F3′5′H, flavonoid 3′5′-hydroxylase; LDOX, leucoanthocyanidin dioxygenase; UF3GT, UDP-glucose:flavonoid-3-*O*-glucosyltransferase

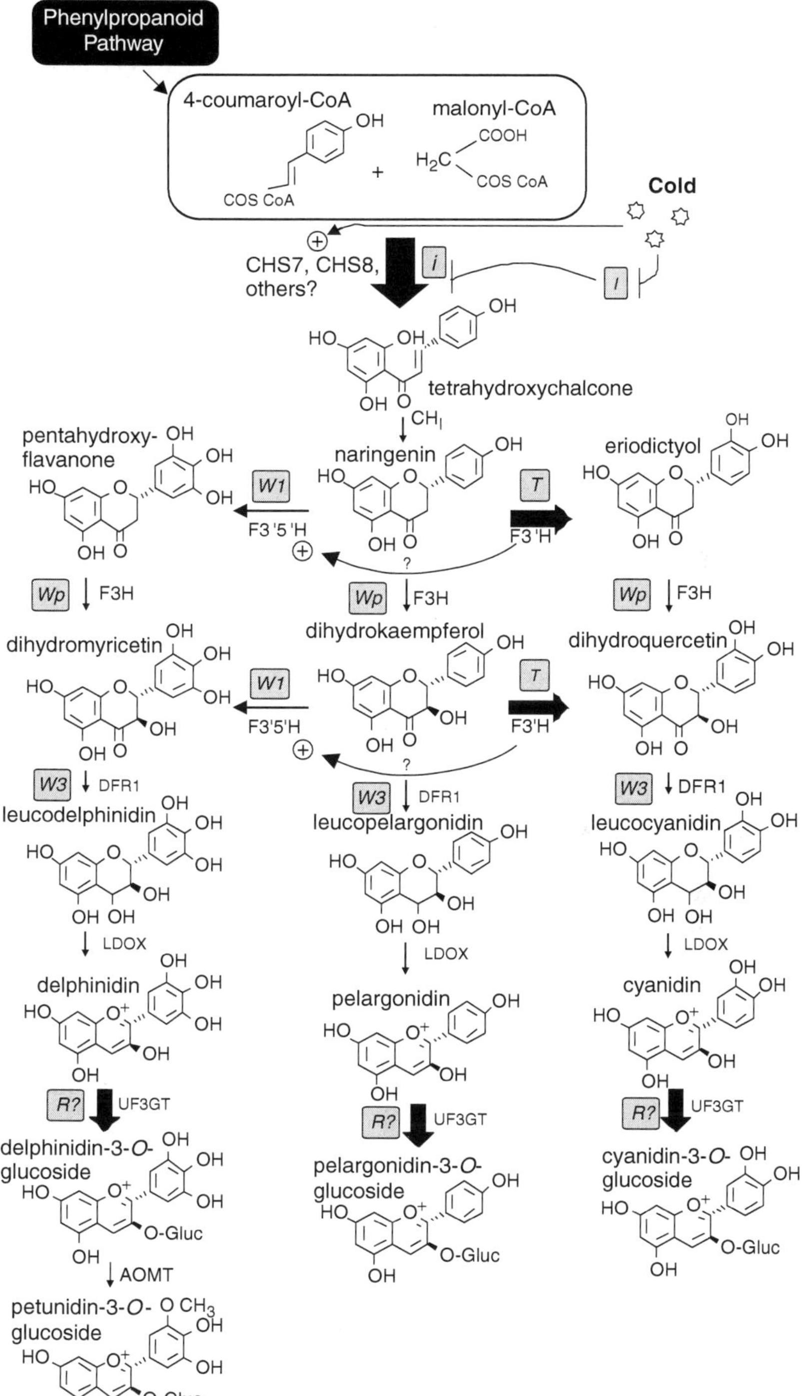

Fig. 4.1 (continued)

(Fig. 4.1). Allelic differences in the *I* locus were attributed to sequence differences in their promoters [15]; however, the transcription factors (TFs) that interact with these promoters have not been identified. In addition, environmental factors such as low-temperature stress appear to affect *I* locus and *CHS* expression [16] and represent an additional level of regulation that remains largely unexplored (Fig. 4.1).

The *T* locus encodes the enzyme flavonoid 3′-hydroxylase (F3′H) [17, 18], and is an important controller of flux in the anthocyanin pathway in soybean seed coats (Fig. 4.1). F3′H diverts metabolic flux away from biosynthesis of orange (pelargonidin) and blue (delphinidin) anthocyanins toward the red cyanidin-3-*O*-glucoside, which is the main anthocyanin in the seed coats of black soybean [7, 8]. *T* increases the accumulation of delphidin-3-*O*-glucoside in black seed coats, even though it is not required for its biosynthesis [19]. Possible mechanisms for this include positive feedback, or the stabilization of the putative anthocyanin biosynthetic metabolon [20] by F3′H-derived membrane anchoring (Fig. 4.1).

The *W1* locus encodes flavonoid 3′5′-hydroxylase (*F3′5′H*) [21]. F3′5′H diverts metabolic flux into the blue delphinidin branch of anthocyanin biosynthesis (Fig. 4.1). In the absence of F3′H activity (*t*), *W1* and recessive *w1* give imperfect black and buff seed colors, respectively [10]. However, in black seeds, *F3′H* (*T*) phenotypically masks *W1*. In contrast to its role in seeds, *W1* has a prominent role in flower colors, as delphinidin-based anthocyanins are the major pigments in purple soybean flowers [22, 23]. Interestingly, *F3′5′H* is expressed at very low levels in flowers and seeds [21]. This suggests that, out of the two branch-point genes (i.e., *F3′H* and *F3′5′H*), it is the strong expression of *F3′H* in seed coats and weak expression in the flowers that determines preferential accumulation of cyanidin-based and delphinidin-based anthocyanins in these respective tissues [21].

In contrast to *W1*, the recessive allele of the *Wp* locus (*wp*) results in a change from black to a lighter grayish seed coat color [24]. *Wp* encodes flavanone 3-hydroxylase (*F3H1*) [24] an enzyme essential for anthocyanin biosynthesis (Fig. 4.1). Similar to *F3′H*, *F3H1* is expressed at high levels in the seed coat relative to the flowers. A reduction in the expression of *F3H1* in the seed coat by the wp mutation may not be sufficient to cause a drastic visible change in seed coat color phenotype (black to gray) as it does in the flower (purple to pink).

The *R* locus determines the presence (*R*) or absence (*r*) of anthocyanins in the seed coat. *R* is required (with *i* and *T*) to produce black seed [10]. However the identity of the gene product encoded by this locus has not been reported. Todd and Vodkin [25] have demonstrated that brown seed coats (*r*) contain proanthocyanidin (PAs) and black seed coats (*R*) contain anthocyanins in addition to PAs and suggested that *R* acts subsequent to the formation of leucoanthocyanidin but previous to the formation of anthocyanins. UDP-glucose:flavonoid 3-*O*-glucosyltransferase (*UF3GT*) should be considered a candidate gene of the *R* locus but its identification has not yet been reported.

Perhaps the least understood of the five loci that control seed coat color in soybean is the *O* locus [10, 11]. The *O* locus only affects the phenotype of brown

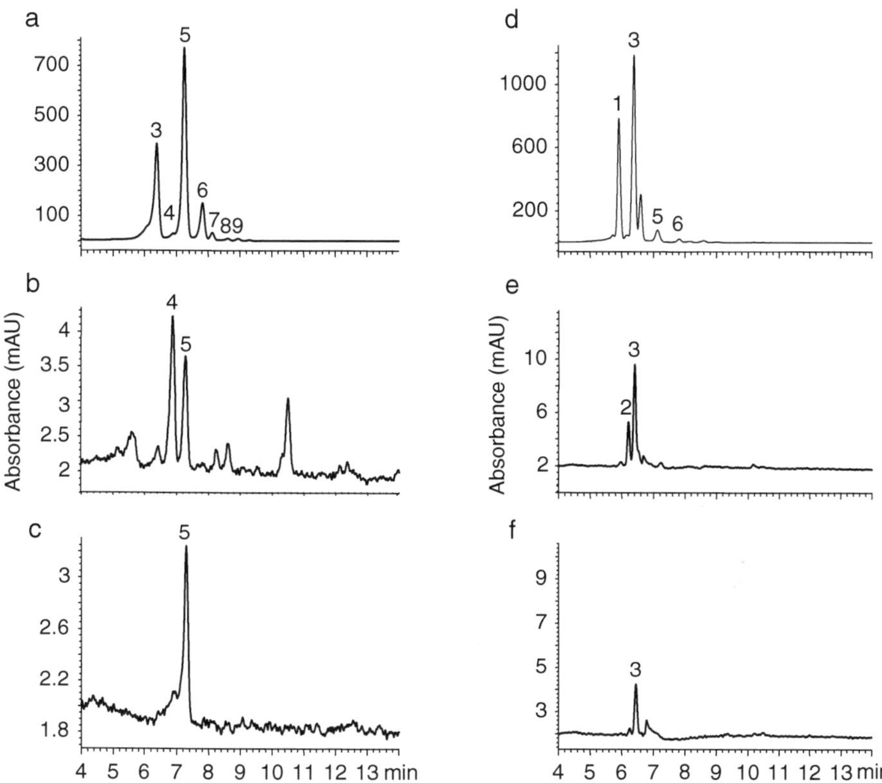

Fig. 4.2 HPLC-DAD chromatograms of anthocyanins from soluble and insoluble extracts of black, red-brown, and brown soybean seed coats at 520 nm. (**a**, **d**) Black (Clark), (**b**, **e**) red-brown (M11), and (**c**, **f**) brown (M100) seed coats. (**a** and **b**) Pulverized fresh seed coats extracted with 80% methanol in water. (**d–f**) Insoluble pulverized seed coat fraction extracted with 1-butanol/HCl (19:1) 1% SDS. Compound identifications were based on comparison of retention times and absorption spectra to authentic standards. Peak 1, unknown; peak 2, delphinidin-3-*O*-galactoside; peak 3, delphinidin-3-*O*-glucoside; peak 4, cyanidin-3-*O*-galactoside; peak 5, cyanidin-3-*O*-glucoside; peak 6, petunidin-3-*O*-glucoside; peak 7, pelargonidin-3-*O*-glucoside; peak 8, peonidin-3-*O*-glucoside; and peak 9, malvidin-3-*O*-glucoside

(*I T r*) seeds [10]. *O* gives brown, whereas recessive *o* gives a red-brown phenotype. Preliminary analysis reveals the anthocyanin content of red-brown seed coats to differ from brown by the presence of 3-*O*-galactosides of cyanidin and delphinidin (Fig. 4.2). An enzyme with minor 3-*O*-galactosyltransferase activity has been identified [26], but it remains to be determined whether genetic differences in this gene account for red-brown versus brown seed coat color.

Four additional genetic loci (W2, W3, W4, and Wm) affect flower but not seed coat color and have been summarized elsewhere [27].

4.3 Is the Black Color in Plants Determined by Anthocyanins?

4.3.1 Anthocyanin Levels Versus Composition in Black Plant Tissues

It is currently not understood whether specific anthocyanin types, levels, or additional factors are required to color plant tissues black. Cyanidin monoglycosides are the predominant anthocyanins in black rice [28] and seven species of black berries [29, 30]. By contrast, delphinidin monoglycosides are the major anthocyanins in black *Lisianthius nigrescens* flowers [31], black tulip [32], and the seed coats of black bean (*Phaseolus vulgaris* L.) [33]. Black carrot [34] and black currant [30] contain anthocyanin 3-diglycosides, but black pansy petals are perhaps the only reported plant organ to contain major anthocyanins glycosylated at the 5- and 3-positions [35]. Glycosides of other anthocyanidins have not been reported to constitute the major anthocyanins of black plant tissues. Cyanidin-3-*O*-glucoside is the most abundant anthocyanin in black soybean [7, 8], whereas the 3-*O*-galactoside is the most abundant in red-brown soybeans (Fig. 4.2). Relative to black soybean, red-brown seed coats have a simple anthocyanin composition and reduced anthocyanin levels (Fig. 4.2).

Anthocyanin levels are significantly higher in black fruits versus their red counterparts. Data from Wu et al. [30] show that black plum and black currant have anthocyanin levels 6 and 36 times higher than their respective red counterparts. Presently, there have been no reports of engineering black color in plant tissues despite numerous successful attempts to increase anthocyanin levels by overexpressing structural and regulatory genes. A 350-fold increase in anthocyanin levels in tomatoes by overexpression of two snapdragon transcription factors resulted in dark purple rather than "black" color [36]. This suggests one or more additional factors may be involved in coloring tissues black.

4.3.2 Are Anthocyanic Vacuolar Inclusions (AVIs) Required to Color Plant Tissues Black?

The extraordinary accumulation of anthocyanins in black petals of wild *L. nigrescens* is thought to be achieved in part by the formation of anthocyanic vacuolar inclusions (AVIs) [31]. Their uncommon presence in flower petals has been shown to enhance both color darkness and intensity [37, 38]. AVIs are insoluble subvacuolar bodies consisting of a protein matrix that binds specific anthocyanins, concentrating them to levels higher than found freely in vacuolar solution [38]. In the AVIs of lisianthius, minor acylated anthocyanins were bound to a protein matrix. Delphinidin triglycosides, with a terminal rhamnose, were "trapped" by this structure, whereas those that lacked this moiety were located in the surrounding vacuolar solution [38]. Pourcel et al. [39] have recently demonstrated that a change in anthocyanin content from 3,5-glycosylated to only 3-glycosylated molecules increases

Fig. 4.3 Bright field microscopy of palisade peels from soybeans at early maturation stage (400–500 mg fresh seed weight). (**a**) Black (Clark) with AVIs (black arrowheads). (**b**) Red-brown (M11). Scale bar = 20 μm. Effect of oxidation on seed coat color demonstrated by exposing immature seeds (400–500 mg fresh seed weight) (**c–f**) to air for 2 h (**g–j**). (**c, g**) Black (Clark), (**d, h**) red-brown (M11), (**e, i**) brown (M100), and (**f, j**) yellow (Clark)

AVI frequency in Arabidopsis. These reports suggest that anthocyanin conjugation may dictate how anthocyanins are packaged. The seed coat palisade of black soybean has vesicle-like AVIs (Fig. 4.3a), similar to those found in the abaxial epidermis of *L. Nigrescens*' petals [40]. Interestingly, the seed coat palisade cells of red-brown soybean do not contain AVIs (Fig. 4.3b). Soluble anthocyanin profiles from the black soybean seed coat (Fig. 4.2a) were similar to those reported previously [8]. Cyanidin-3-*O*-glucoside (Fig. 4.2a, peak 5) was the major anthocyanin, but the 3-*O*-galactoside (Fig. 4.2b, peak 4) was predominant in red-brown seed coats. By contrast, more stringent extraction of the insoluble seed coat fraction resulted in completely different anthocyanin profiles (Fig. 4.2). Both black (Fig. 4.2d, peak 3) and red-brown (Fig. 4.2e, peak 3) soybean seed coats contained delphinidin 3-*O*-glucoside as the main insoluble anthocyanin, but black seed coats contained an additional unknown compound (Fig. 4.2d, peak 1). Whether high levels of delphinidin 3-*O*-glucoside or the unknown compound is required for AVI formation and black seed coat color in soybean seeds remains to be determined.

4.3.3 Does Co-pigmentation and Oxidation Have a Role in Blackening Seeds?

The role of co-pigmentation in black tissues may be best exemplified by the flowers of wild *L. nigrescens*, which accumulate an equal ratio of anthocyanins and flavones [31]. Co-pigmentation is generally considered to occur by physical interactions between anthocyanins and other molecules [41]. In brown seeds such as brown soybean, oxidized proanthocyanidins (PAs) are the major pigments [42]. PAs are flavonoid polymers or oligomers composed of (epi)flavan-3-ol units. The polymerization reaction of PAs is not yet fully understood but may require enzymatic and/or nonenzymatic oxidation reactions [43–45]. In black and brown soybean seed coats, proanthocyanidin molecules are relatively large (up to 30 subunits [45]), suggesting frequent oxidative reactions. Anthocyanin and PA accumulation occurs simultaneously in the seed coat palisade of black soybean at later stages of development [26] and both molecules generally accumulate in the vacuole [42]. To determine whether oxidation may have a role in producing black soybean seed color, fully grown seeds were exposed to air to mimic the effect of oxidation that occurs during desiccation (Fig. 4.3c–j). Anthocyanin- and PA-containing seeds of black soybean oxidized to black (Fig. 4.3c, g) while seeds that contained only PAs browned (Fig. 4.3e, i) and immature red-brown seeds that contain PAs and lower levels of anthocyanins relative to black darkened (Fig. 4.3d, h). Seeds that lacked both pigments (except in the hilum) did not change significantly (Fig. 4.3f, j). This suggests that oxidation in air has a role in enhancing the coloration of black soybeans. Reducing PA levels may be required to engineer vibrant seed colors for the visual identification of transgenic grains.

4.4 Engineering Seed Coat Color for the Visual Identification of Transgenic Grains

Coloring transgenic plants, plant organs, and other transgenic materials intended for commercial markets to enable their visual identification could provide increased safety features to consumers and increased freedom to producers. These have been cited as the two main issues that threaten the adoption of genetically engineered plants and foods [1]. The introduction of foreign proteins or metabolites could potentially be used to mark transgenic materials. However, the introduction of a substance that is new to the food supply would require additional regulatory scrutiny under the concept of substantial equivalence [47], as it may pose increased health or environmental risks. Some potentially safe alternatives for labeling transgenic materials, such as using green fluorescent protein [48], may not be practical, as machinery are required for their detection. A potentially safe and economical alternative would be to produce distinct colors in transgenic plants of natural products that are currently already part of the food supply and that are generally recognized as safe. To exemplify this concept, we are presently attempting to suppress anthocyanin and/or proanthocyanidin (PA) biosynthesis in the seed coats of black soybean. We have chosen to suppress genes specific to the anthocyanin and PA pathways to minimize the potential for unintended effects on food or feed biosafety.

4.5 Concluding Remarks and Future Perspectives

Recent advances have improved our understanding of anthocyanin biosynthesis, composition, and flux in black soybean. However, several areas such as regulatory genes, environment-responsive signaling pathways, and mechanisms of AVI formation remain largely uninvestigated. Our understanding of black color in plant tissues is presently speculative, but future efforts to engineer seed color in black soybean will provide great insight. Engineering reduced anthocyanin levels in the seed coat of black soybean using RNAi or anti-sense technologies could potentially be used to produce distinct seed colors to enable the visual identification of transgenic grains. Genes encoding herbicide tolerance, pest resistance, or other commercial traits could be genetically linked to color-modifying elements in a single construct for plant transformation. Coloring transgenic materials to enable their visual identification could provide increased safety to consumers and freedom to producers and should not be limited in the future to anthocyanin biosynthesis or black soybean.

Acknowledgments The authors thank Dr. Ammar Saleem (University of Ottawa) for his assistance with HPLC, Dr. Shea Miller for performing the microscopy, and Drs. Malcolm Morrison and Elroy Cober (Agriculture and Agri-Food Canada) for providing seeds and for their helpful discussions. We would also like to thank the reviewers for their helpful suggestions. The research was supported by an NSERC Discovery Grant and AAFC project (RBPI 126) to BM.

References

1. Lemaux PG (2008) Genetically engineered plants and foods: a scientist's analysis of the issues (Part I). Annu Rev Plant Biol 59:771–812
2. Doyle A (2008) EMBARGOED – World fails to monitor biotech trade – UN study. Williams R (ed). Thomson Reuters
3. Meyer P, Heidmann I, Forkmann G et al (1987) A new petunia flower colour generated by transformation of a mutant with a maize gene. Nature 330:677–678
4. Tanaka Y, Katsumoto Y, Brugliera F et al (2005) Genetic engineering in floriculture. Plant Cell, Tiss and Org Cult 80:1–24
5. Katsumoto Y, Fukuchi-Mizutani M, Fukui Y et al (2007) Engineering of the rose flavonoid biosynthetic pathway successfully generated blue-hued flowers accumulating delphinidin. Plant Cell Physiol 48:1589–1600
6. Holton T, Tanaka Y (1994) Blue roses-a pigment of our imagination? Trends Biotechnol 12:40–42
7. Choung M, Baek I-Y, Kang S-T et al (2001) Isolation and determination of anthocyanins in seed coats of black soybean (Glycine max (L.) Merr.). J Agric Food Chem 49:5848–5851
8. Lee JH, Kang NS, Shin S-O et al (2009) Characterization of anthocyanins in the black soybean (Glycine max L.) by HPLC-DAD-ESI/MS analysis. Food Chem 112:226–231
9. Springob K, Nakajima J, Yamazaki M et al (2003) Recent advances in the biosynthesis and accumulation of anthocyanins. Nat Prod Rep 20:288–303

10. Bernard RL, Weiss MG (1973) Qualitative genetics. In: Caldwell BE (ed) Soybeans: improvement, production, and uses, 1st edn. Am Soc Agron, Wisconsin

11. Palmer RG, Kilen TC (1987) Qualitative genetics and cytogenetics. In: Wilcox JR (ed) Soybeans: improvement, production and uses, 1st edn. Am Soc Agron, Wisconsin

12. Tuteja JH, Vodkin LO (2008) Structural features of the endogenous *CHS* silencing and target loci in the soybean genome. Crop Sci 48:S49–S68

13. Senda M, Masuta C, Ohnishi S et al (2004) Patterning of virus-infected *Glycine max* seed coat is associated with suppression of endogenous silencing of chalcone synthase genes. Plant Cell 16:807–818

14. Tuteja JH, Clough SJ, Chan WC et al (2004) Tissue-specific gene silencing mediated by a naturally occurring chalcone synthase gene cluster in Glycine max. Plant Cell 16:819–835

15. Kasai A, Ohnishi S, Yamazaki H et al (2009) Molecular mechanism of seed coat discoloration induced by low temperature in yellow soybean. Plant Cell Physiol 50:1090–1098

16. Toda K, Yang D, Yamanaka N et al (2002) A single-base deletion in soybean flavonoid 3′-hydroxylase gene is associated with gray pubescence color. Plant Mol Biol 50:187–196

17. Zabala G, Vodkin L (2003) Cloning of the pleiotropic T locus in soybean and two recessive alleles that differentially affect structure and expression of the encoded flavonoid 3′ hydroxylase. Genetics 163:295–309

18. Buzzell R, Buttery B, MacTavish D (1987) Biochemical genetics of black pigmentation of soybean seed. The J Hered 78:53–54

19. Saslowsky D, Winkel-Shirley B (2001) Localization of flavonoid enzymes in Arabidopsis roots. Plant J 27:37–48

20. Winkel BSJ (2009) Metabolite channeling and multi-enzyme complexes. In: Osbourn AE, Lanzotti V (eds) Plant-derived natural products: synthesis, function, and application. Springer, New York

21. Zabala G, Vodkin LO (2007) A rearrangement resulting in small tandem repeats in the F3′5′H gene of white flower genotypes is associated with the soybean W1 locus. Plant Genome 47:S113–S124

22. Iwashina T, Githiri SM, Benitez ER et al (2007) Analysis of flavonoids in flower petals of soybean near-isogenic lines for flower and pubescence color genes. J Hered 98:250–257

23. Iwashina T, Oyoo ME, Khan NA et al (2008) Analysis of flavonoids in flower petals of soybean flower color variants. Crop Sci 48:1918–1924

24. Zabala G, Vodkin LO (2005) The wp mutation of Glycine max carries a gene-fragment-rich transposon of the CACTA superfamily. Plant Cell 17:2619–2632

25. Todd JJ, Vodkin LO (1993) Pigmented soybean (Glycine max) seed coats accumulate proanthocyanidins during development. Plant Physiol 102:663–670

26. Kovinich N, Saleem A, Arnason JT, Miki B (2010) Functional characterization of a UDP-glucose:flavonoid 3-O-glucosyltransferase from the seed coat of black soybean (Glycine max (L.) Merr). Phytochemistry 71:1253–1263

27. Takahashi R, Matsumura H, Oyoo ME et al (2008) Genetic and linkage analysis of purpleblue flower in soybean. J Hered 99:593–597

28. Ryu SN, Park SZ, Ho C-T (1998) High performance liquid chromatographic determination of anthocyanin pigments in some varieties of black rice. J Food Drug Anal 6:729–736

29. Stintzing FC, Stintzing AS, Carle R et al (2002) A novel zwitterionic anthocyanin from evergreen blackberry (Rubus laciniatus Willd). J Agric Food Chem 50:396–399

30. Wu X, Beecher GR, Holden JM et al (2006) Concentrations of anthocyanins in common foods in the United States and estimation of normal consumption. J Agric Food Chem 54:4069–4075

31. Markham KR, Bloor SJ, Nicholson R et al (2004) Black flower coloration in wild *Lisianthius nigrescens*: its chemistry and ecological consequences. Z Naturforsch C 59:625–630

32. Shibata M, Ishikura N (1960) Paper chromatographic survey of anthocyanin in tulipflowers. I Jap J Bot 17:230–238

33. Takeoka GR, Dao LT, Full GH et al (1997) Characterization of black bean (*Phaseolus vulgaris* L.) anthocyanins. J Agric Food Chem 45:3395–3400

34. Stintzing FC, Stintzing AS, Carle R et al (2002) Color and antioxidant properties of cyanidin-based anthocyanin pigments. J Agric Food Chem 50:6172–6181
35. Goto T, Takase S, Kondo T (1978) PMR spectra of natural acylated anthocyanins. Determination of the stereostructure of awobanin, shisonin and violanin. Tetrahedron Lett 27:2413–2416
36. Butelli E, Titta L, Giorgio M et al (2008) Enrichment of tomato fruit with health-promoting anthocyanins by expression of select transcription factors. Nat Biotechnol 26:1301–1308
37. Gonnet JF (2003) Origin of the color of Cv. rhapsody in blue rose and some other so-called "blue" roses. J Agric Food Chem 51:4990–4994
38. Markham KR, Gould KS, Winefield CS et al (2000) Anthocyanic vacuolar inclusions – their nature and significance in flower colouration. Phytochem 55:327–336
39. Pourcel L, Irani NG, Lu Y et al (2010) The formation of anthocyanic vacuolar inclusions in arabidopsis thaliana and implications for the sequestration of anthocyanin pigments. Molec Plant 3(1):78–90
40. Zhang H, Wang L, Deroles S et al (2006) New insight into the structures and formation of anthocyanic vacuolar inclusions in flower petals. BMC Plant Biol 6:29
41. Brouillard R, Dangles O (1994) Flavonoids and flower colour. In: Harborne JB (ed) The flavonoids-advances in research since 1986. CRC Press, Boca Raton, FL
42. Lepiniec L, Debeaujon I, Routaboul JM et al (2006) Genetics and biochemistry of seed flavonoids. Annu Rev Plant Biol 57:405–430
43. Dixon RA, Xie DY, Sharma SB (2005) Proanthocyanidins – a final frontier in flavonoid research? New Phytol 165:9–28
44. Pourcel L, Routaboul JM, Cheynier V et al (2007) Flavonoid oxidation in plants: from biochemical properties to physiological functions. Trends Plant Sci 12:29–36
45. Pourcel L, Routaboul JM, Kerhoas L et al (2005) TRANSPARENT TESTA10 encodes a laccase-like enzyme involved in oxidative polymerization of flavonoids in Arabidopsis seed coat. Plant Cell 17:2966–2980
46. Takahata Y, Ohnishi-Kameyama M, Furuta S et al (2001) Highly polymerized procyanidins in brown soybean seed coat with a high radical-scavenging activity. J Agric Food Chem 49:5843–5847
47. OECD (1993) Safety evaluation of foods derived by modern biotechnology, concepts and principles. Org Econ Coop Dev, Paris
48. Richards HA, Han CT, Hopkins RG et al (2003) Safety assessment of recombinant green fluorescent protein orally administered to weaned rats. J Nutr 133:1909–1912

Chapter 5
Pharmacogenetics in Potential Herb–Drug Interactions: Effects of Ginseng on CYP3A4 and CYP2C9 Allelic Variants

Alice Luu, Brian C. Foster, Kristina L. McIntyre, Teresa W. Tam, and John T. Arnason

5.1 Introduction: Ginseng Drug Interactions

Ginseng species (*Panax* spp.) have been used for thousands of years to treat a variety of ailments including fatigue, insomnia, impotence, and chronic illnesses such as diabetes. There are currently 16–18 recognized species of *Panax*, depending on taxonomic treatment [1]. *Panax ginseng* C.A. Meyer (Asian ginseng) and *Panax quinquefolius* L. (American ginseng) are the most commonly used and commercially grown species. Their pharmacological activities have been attributed to several bioactive components, of which the ginsenosides, also known as triterpenoid glycosides, are the most important (Fig. 5.1) [2].

In general, ginseng is not associated with any serious adverse effects if taken at recommended doses. However, there have been frequent reports of adverse reactions such as insomnia, gastric irritation, anxiety, and headaches. There has also been growing concern for the development of herb–drug interactions with the use of ginseng. Recently, Health Canada has indicated cautions for the concomitant use of *P. ginseng* with warfarin and phenelzine [3, 4]. Although other studies have not been able to confirm such interactions, these reports suggest that the administration of ginseng may interfere with the metabolism of drugs in the body.

5.2 Cytochrome P450 (CYP) Enzymes

CYP3A4 and CYP2C9 are the major drug-metabolizing enzymes involved in the biotransformation of over 70% of clinically prescribed drugs. Ginseng could potentially inhibit or induce these enzymes, predisposing an individual to potential herb–drug interactions. Indeed, some studies in the literature have shown that ginsenosides, along with their metabolites, can inhibit CYP3A4 and CYP2C9 [5–8].

B.C. Foster (✉)
Department of Cellular and Molecular Medicine, University of Ottawa, Ottawa, ON, Canada;
Therapeutic Products Directorate, Health Canada, Ottawa, ON, Canada
e-mail: brian_foster@hc-sc.gc.ca

D.R. Gang (ed.), *The Biological Activity of Phytochemicals*, Recent Advances in Phytochemistry 41, DOI 10.1007/978-1-4419-7299-6_5,
© Springer Science+Business Media, LLC 2011

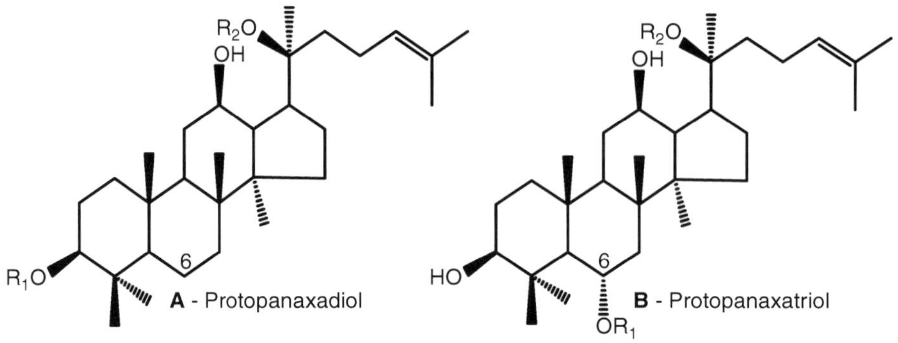

R_2O
OH

A - Protopanaxadiol

R_1O

6

R_2O
OH

B - Protopanaxatriol

HO

6

OR_1

	Ginsenoside	R_1	R_2
A	Rb$_1$	-glucose2-glucose	-glucose6-glucose
	Rb$_2$	-glucose2-glucose	-glucose6-arabinose (pyranose)
	Rc	-glucose2-glucose	-glucose6-arabinose (furanose)
	Rd	-glucose2-glucose	-glucose
B	Re	-glucose2-rhamnose	-glucose
	Rg$_1$	-glucose	-glucose

Fig. 5.1 Chemical structures of ginsenosides. Ginsenosides can be separated into protopanaxadiols (**a**) or protopanaxatriols (**b**), depending on the presence or absence of a substituent group at carbon 6. Individual ginsenosides differ according to their sugar moieties, as indicated by R_1 and R_2 in the table

Other studies, however, have not been able to reproduce these results. Etheridge et al. [9], for example, did not find any significant alteration of enzyme activity using a ginseng root extract and ginsenoside metabolites in human liver microsomes (HLM) containing CYP3A4 and CYP2C9. Based on the totality of evidence, it remains unclear whether the administration of ginseng could have a significant or physiologically relevant effect on CYP3A4 and CYP2C9 activity.

5.3 The Role of Polymorphisms

Pharmacogenetics is another important factor to consider during the assessment of herb–drug interactions. In the past, herb–drug interactions have only been examined using wild-type CYP enzymes with normal function. However, polymorphisms in the human genome are a common occurrence, leading to modified enzyme structure and/or function. Indeed, many polymorphisms have been identified in CYP genes, including those encoding CYP3A4 and CYP2C9. Interestingly, the *CYP3A4* gene can tolerate many polymorphisms in the genetic code without being compromised in function. The gene for *CYP2C9*, however, is not as tolerant and polymorphisms often give rise to mutant alleles associated with clinical complications [10]. Although over 50 polymorphisms have been identified in the *CYP2C9* gene, only two allelic variants (CYP2C9*2 and CYP2C9*3) have been considered to be clinically relevant and both variants harbor a single nucleotide mutation [11]. These alleles are most prominent in Caucasian populations and are associated with a decrease in enzyme activity when compared to the wild-type allele. Although research has examined the impact of CYP2C9 polymorphisms on drug metabolism, little is known about the potential of serious interactions when CYP2C9*2 or CYP2C9*3 is inhibited. Furthermore, there is currently no information on the effects of ginseng, one of the most widely used herbs, on CYP2C9 allelic variants.

5.4 Recent Results on CYP3A4 and CYP2C9 Allelic Variants

The effects of ginseng on CYP3A4, CYP2C9*1, CYP2C9*2, and CYP2C9*3 were recently examined in our laboratory using a method adapted from Tam et al. [12]. A collection of 6–8 ginseng products was tested in 96-well plates, where activity was detected using a fluorescence-based plate reader. The four commercial root products were assigned Natural Product Number (NPN) accession numbers and vouchers have been deposited in the herbarium at the University of Ottawa (U of O). The in-house preparations consisted of one *P. ginseng* C.A. Meyer sample and three *P. quinquefolius* L. root samples while the commercial products were root samples equally representative of both species. To maintain company anonymity, the commercial products from local distributors were coded NPN1, NPN2 for *P. ginseng* samples and NPN3, NPN4 for *P. quinquefolius* samples. In brief, all ginseng products were prepared as extracts and incubated with CYP enzymes, substrate (dibenzylfluorescein, DBF; 7-methyloxytrifluorocoumarin, MFC), nicotinamide adenine dinucleotide phosphate, reduced form (NADPH), and reaction buffer [12]. In addition to these bioassays, the ginseng products were also analyzed using high-performance liquid chromatography with diode array detection (HPLC-DAD) to characterize them for six major ginsenosides (Rg_1, Re, Rb_1, Rc, Rb_2, and Rd) [13].

Our results showed that all ginseng products had weak (0–40%) to moderate (40–80%) inhibitory activity, depending on the CYP enzyme studied. For CYP3A4, it was found that the ethanol (EtOH), aqueous, and methanol (MeOH) extracts of Ontario ginseng prepared in-house were significantly more inhibitory than any other tested product (Fig. 5.2). Based on these results, we tested whether the degree of CYP3A4 inhibition by different ginseng products could be correlated with total ginsenoside content (% w/w). The correlation between inhibitory activity and ginsenoside content was statistically significant with a very high r^2 value (Fig. 5.3, Table 5.1).

When the three allelic variants of CYP2C9 were studied, the commercial products inhibited the wild-type enzyme (CYP2C9*1) to a comparable degree as CYP3A4, but the Ontario ginseng samples showed much less inhibition. The CYP2C9*3 variant gave a profile of inhibition similar to the wild-type enzyme, but there was low inhibition of CYP2C9*2 by all products. When inhibition of each of

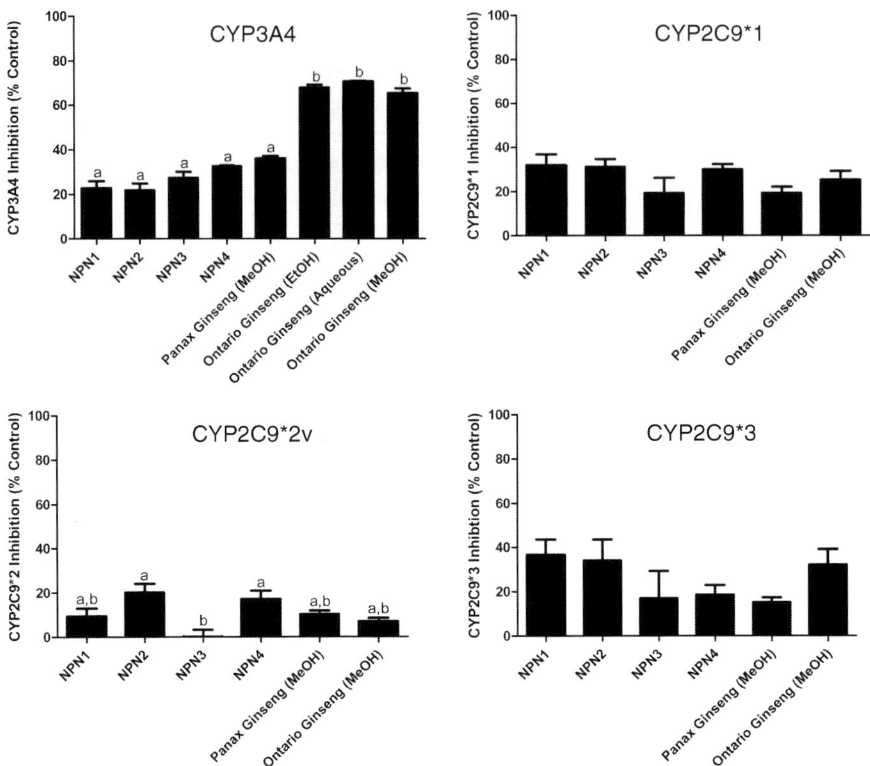

Fig. 5.2 The inhibitory potential of different ginseng products on CYP3A4 and three CYP2C9 allelic variants. A typical 200 μL reaction mixture contained cDNA-expressed enzymes (0.3 pmol for CYP3A4, 5 pmol for CYP2C9*1, 5 pmol for CYP2C9*2, and 10 pmol for CYP2C9*3), NADPH (0.54 mM), probe substrates (1 μM dibenzylfluorescein for CYP3A4, 0.2 mM MFC for CYP2C9 assays), and ginseng samples (80 μg/mL) in 185 mM reaction buffer. Except for CYP2C9*3 whose incubation time was set to 60 min, all other CYP assays were incubated for 20 min at 37°C. All data were expressed as means ± SEM. Samples were tested in triplicate with assays repeated three times ($n = 3$). One-way ANOVA was performed, followed by Bonferroni multiple comparison tests. Means with significant differences are distinguished by different letters ($p<0.05$)

the three CYP2C9 variants was correlated with ginsenoside content, no significant correlation was found. In summary, the present *in vitro* findings showed that the inhibition of CYP3A4 was dependent on ginsenoside concentration across all products, but this was not observed for any of the allelic variants of CYP2C9. Ontario-grown ginseng had the highest inhibitory effect on CYP3A4 due to its high ginsenoside content. Such interactions could cause the plasma levels of concomitantly administered drugs to be higher, posing a potential risk for serious herb–drug interactions. While the *in vitro* results predict a possible interaction, further clinical studies are needed to determine whether the effect can occur *in vivo*.

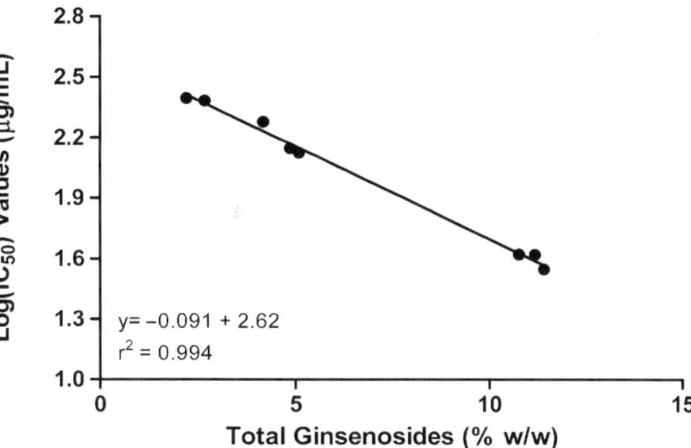

Fig. 5.3 Correlation between total ginsenoside content and log-transformed IC_{50} values. The IC_{50} values of different ginseng products were plotted against their total ginsenoside content. Only products quantified by HPLC-DAD were included in the analysis. Pearson correlation yielded a value of $r = -0.997$ ($p<0.0001$)

Table 5.1 Characterization of ginsenoside content and composition across products through HPLC-DAD analysis

Ginseng product[a]	Ginsenoside profile						
	Rg_1 (%)	Re (%)	Rb_1 (%)	Rc (%)	Rb_2 (%)	Rd (%)	Total (%)
NPN3	0.160	1.04	2.33	0.258	0.059	0.324	4.18
NPN4	0.201	0.857	2.30	0.670	0.120	0.712	4.86
Panax Ginseng (MeOH)	1.24	0.355	1.73	1.00	0.576	0.184	5.09
Ontario ginseng (EtOH)	0.509	3.26	5.16	0.840	0.187	0.793	10.7
Ontario ginseng (Aqueous)	0.542	3.37	5.79	0.803	0.199	0.687	11.4
Ontario ginseng (MeOH)	0.160	1.86	5.92	0.814	0.156	2.24	11.1

[a]Only products with chromatograms suitable for quantification were presented. Owing to the presence of excipients, NPN1 and NPN2 could not be quantified using HPLC-DAD. These products were omitted in subsequent statistical comparisons and correlations.

In terms of the impact of polymorphisms, our results demonstrated that genetic mutations did not lead to differences in the inhibition of CYP2C9 or its allelic variants. This suggests that genetic background may not be involved in the predisposition of herb–drug interactions for ginseng with this enzyme.

5.5 Future Implications

Herb–drug interactions are of growing concern due to the increased use and awareness of natural health products. They generally arise when natural health products inhibit CYP enzymes, altering the rate of metabolism for other drugs in the system. It is important to note, however, that interactions may also arise when

CYP enzymes are induced, resulting in reduced plasma drug levels. Alternatively, CYP enzymes could also undergo mechanism-based inhibition, whereby a CYP enzyme can be completely inactivated by covalent bonding to a component of the herb. Furthermore, botanicals can elicit a biphasic cellular response, whereby CYP activity may be inhibited initially, followed by induction after prolonged incubation or repeated administration. Such factors would need to be considered in future studies in order to establish the true risk of ginseng in herb–drug interactions.

Although the results from the present study suggest that the risk of herb–drug interactions is low even in the presence of polymorphisms, extrapolating *in vitro* findings to clinical settings is difficult due to a number of extraneous factors that need to be considered. For example, our study only investigated the effect of ginseng products on human cDNA-expressed enzymes with a single substrate. In clinical practice, different CYP2C9 genotypes including CYP2C9*1/*2, CYP2C9*1/*3, CYP2C9*2/*3 with varying amounts of probe compounds would need to be considered before drawing conclusions. Furthermore, these genotypes only represent a limited subset of all possible combinations, and are associated with different pharmacokinetic profiles. Such complex relationships have been a common challenge in the field of pharmacogenetics and further research would be required to establish the true clinical relevance of polymorphisms in herb–drug interactions.

5.6 Conclusions

Natural health product consumers are generally under the impression that alternative treatments are safe and efficacious since they are derived from natural sources. Furthermore, many of these consumers do not consult with their physicians while taking medications. This practice can potentially lead to severe clinical complications if not monitored carefully. The risk of developing herb–drug interactions is affected by many factors, including the quality and species of the herb used, the concentration of active CYP enzyme inhibitors, as well as the intrinsic and extrinsic factors that affect drug disposition in an individual. The significance of the particular drug-metabolizing enzyme studied and ginsenoside profile is now apparent and necessary, especially since scientific advancement has made it possible to analyze these factors. It is important for future research to begin examining the impact of genetic polymorphisms on drug metabolism and how such differences could lead to herb–drug interactions. The results of such studies will ultimately help predict and prevent serious herb–drug interactions for natural health product users.

Acknowledgments Extraction and HPLC-DAD methods were kindly validated and provided by Paula Brown (NHP Research Group, British Columbia Institute of Technology, Burnaby, Canada). This project was funded by the Ontario Ginseng Innovation and Research Centre (OGIRC).

References

1. Sharma SK, Pandit MK (2009) A new species of *Panax* from Sikkim Himalaya, India. Syst Bot 34:434–438
2. Bahrke MS, Morgan WP (1994) Evaluation of the ergogenic properties of ginseng. Sports Med 18:229–248
3. Janetzki K, Morreale AP (1997) Probable interaction between warfarin and ginseng. Am J Health Syst Pharm 54:692–693
4. Shader RI, Greenblatt DJ (1985) Phenelzine and the dream machine – ramblings and reflections. J Clin Psychopharmacol 5:65
5. Henderson GL, Harkey MR, Gershwin ME et al (1999) Effects of ginseng components on cDNA-expressed cytochrome P450 enzyme catalytic activity. Life Sci 65:209–214
6. He N, Edeki T (2004) The inhibitory effects of herbal components on CYP2C9 and CYP3A4 catalytic activities in human liver microsomes. Am J Ther 11:206–212
7. Liu Y, Zhang JW, Li W et al (2006) Ginsenoside metabolites, rather than naturally occurring ginsenosides, lead to inhibition of human cytochrome P450 enzymes. J Toxicol Sci 91: 356–364
8. Hao M, Zhao Y, Chen P et al (2008) Structure–activity relationship and substrate-dependent phenomena in effects of ginsenosides on activities of drug-metabolizing P450 enzymes. PloS One 3:e2697
9. Etheridge AS, Black SR, Patel PR et al (2007) An in vitro evaluation of cytochrome P450 inhibition and p-glycoprotein interaction with goldenseal, ginkgo biloba, grape seed, milk thistle, and ginseng extracts and their constituents. Planta Med 73:731–741
10. Aithal GP, Day CP, Kesteven PJL et al (1999) Association of polymorphisms in the cytochrome P450 CYP2C9 with warfarin dose requirement and risk of bleeding complications. Lancet 353:717–719
11. Kirchheiner J, Brockmoller J (2005) Clinical consequences of cytochrome P450 2C9 polymorphisms. Clin Pharmacol Ther 77:1–16
12. Tam TW, Liu R, Arnason JT et al (2009) Actions of ethnobotanically selected Cree antidiabetic plants on human cytochrome P450 isoforms and flavin-containing monooxygenase 3. J Ethnopharmacol 126:119–126
13. Yu R, Brown PN (2008) Single laboratory validation study for the determination of ginsenoside content in *Panax ginseng* C.A. Meyer and *P. quinquefolius L*. Raw Materials and finished Products by High Pressure Liquid Chromatography. 5th Annual NHP Research Conference Toronto Ontario, March 27–29, 2008

Chapter 6
Biosynthesis and Function of Citrus Glycosylated Flavonoids

Daniel K. Owens and Cecilia A. McIntosh

6.1 Citrus Flavonoids and Flavonoid Glycosides

Citrus is one of the long-standing horticultural and agricultural plant genera with a rich history from centuries of cultivation. As a result, there are many species and varieties (cultivars) within species, each with characteristic properties. These properties include vegetative growth pattern as well as fruit characters such as flavor, seediness, size, and appearance.

Flavonoids in fruit and the impact of these compounds on the marketability of fruit and fruit products were the driving forces for much of the early work identifying the major flavonoids in Citrus. In general, Citrus is known for the accumulation of large amounts of flavonoid diglycosides in some fruit and vegetative tissues. The most prevalent diglycoside sugar linkages found are the neohesperidosides and rutinosides. Neohesperidosides are rhamnoglucosides with C-1 of the rhamnose attached to C-2 of the glucose via an O-glycosidic linkage (Fig. 6.1). The 1→2 linkage results in a bitter flavor [1], and one of the more prevalent bitter compounds is the flavanone neohesperidoside, naringin (Fig. 6.1). C-1 of rhamnose is covalently bound to C-6 of the glucose via an O-glycosidic linkage in rutinosides (Fig. 6.1) and flavonoid rutinosides (e.g., narirutin) are tasteless [1]. Some Citrus species such as pummelo (*C. maxima*, *C. grandis*) produce predominantly neohesperidosides, while others, such as orange (*C. sinensis*), produce predominantly rutinosides. Grapefruit (*C. paradisi*) produces both types of 7-O-diglycosides, and this and other evidence support the hypothesis that grapefruit is a result of a cross between pummelo and orange [2, 3].

There have been many survey studies of citrus flavonoids, yet by no means has an exhaustive analysis been performed for all species and cultivars. Surveys often are initiated by hydrolysis of tissue extracts and identification of the flavonoid aglycones present in the tissues. This provides a basis for subsequent identification of the glycosylated compounds present

C.A. McIntosh (✉)
Department of Biological Sciences, East Tennessee State University, Johnson City, TN 37614, USA; School of Graduate Studies, East Tennessee State University, Johnson City, TN 37614, USA
e-mail: mcintosc@etsu.edu

D.R. Gang (ed.), *The Biological Activity of Phytochemicals*, Recent Advances in Phytochemistry 41, DOI 10.1007/978-1-4419-7299-6_6,
© Springer Science+Business Media, LLC 2011

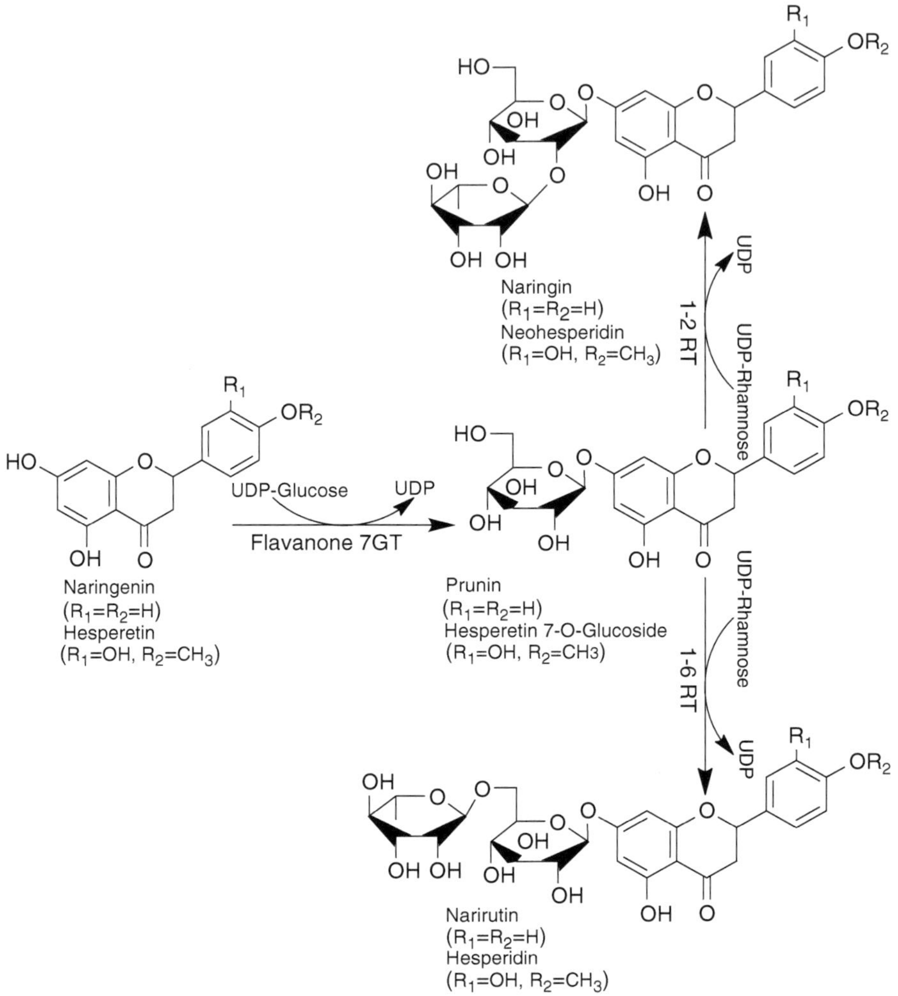

Fig. 6.1 Biosynthesis of naringin, neohesperidin, narirutin, and hesperidin in Citrus

and this has been performed using a variety of strategies. In 2003, the USDA posted a database of flavonoid aglycones in foods (http://www.nal.usda. gov/fnic/foodcomp/Data/Flav/flav.pdf). Information on flavonoid aglycones in fruit can be found at http://www.villagewineryandvineyards.com/USDA-elderberry-Flavonoid-chart.pdf. A relatively comprehensive survey of flavonoids and flavonoid glycosides in many varieties of Citrus was published in 1998 by Berhow et al. [2] including information on tissue source. Peterson et al. [4] reviewed reports of flavanone glycoside content in oranges, mandarins, tangors, tangelos, and sour oranges. A recent review on flavonoids found in fresh-squeezed and commercial citrus juice focusing on sweet orange, tangerine, and hybrids as well as sour orange, lemon, lime, and grapefruit has been published by Gattuso et al. [5]. Additionally, a study of flavonoids found in citrus leaves from plants subject to different environmental conditions has recently been reported [6].

The impact of high levels of production of flavanone and flavone rutinosides and neohesperidosides on consumer acceptance of commercial citrus products has been known for many decades and includes concern of bitterness due to neohesperidoside levels as well as issues of excessive clouding in products from Citrus producing high levels of rutinosides [7, 8, partially reviewed in 5]. Results from studies performed to determine variability in flavonoid glycoside content in different fruit varieties [5, 9–11], levels during fruit development and maturation [12, 13], and during re-differentiation in tissue cultures [14–16] indicate a dynamic metabolic process. These studies and others have renewed interest in the synthesis and the regulation of biosynthesis of flavonoid glycosides in Citrus.

6.2 Biosynthesis of Flavonoids in Citrus

Table 6.1 is a summary of the phenylpropanoid and core flavonoid biosynthetic enzymes that have been established in Citrus, accession numbers for the corresponding genes, and notation of those genes for which identity has been verified through either testing of enzymes isolated from Citrus tissues or heterologously expressed proteins. The following is a review of flavonoid biosynthesis in general with specific information on biosynthesis of core flavonoids in Citrus (Fig. 6.2).

6.2.1 Phenylalanine Ammonia Lyase

A series of radioactive tracer experiments suggested that hydroxycinnamic acids were important predecessors of plant secondary products, including the flavonoids, and it was later proposed by Neish [17] that these compounds could originate from direct deamination of amino acids. Subsequently, the enzyme now known as phenylalanine ammonia lyase (PAL; EC 4.3.1.5) was purified, shown to convert phenylalanine to *trans*-cinnamate, and biochemically characterized from *Hordeum vulgare* L. var. Aravat (barley) [18]. PAL is an important branch point enzyme between primary and secondary metabolism. It appears to be the only connection between the shikimate pathway and the production of flavonoids as well as a number of other natural products. Although tyrosine ammonia-lyase (TAL; 4.3.1.5) was initially proposed to serve as another shikimate pathway connection in monocotyledonous plants, this enzyme has been identified only in a few bacterial species [19]. The TAL activity previously reported in monocots has since been attributed to PAL accepting both phenylalanine and tyrosine as substrates in some plant species, although typically with a stronger preference for phenylalanine.

PAL is a tetramer of 226–340 kDa, and is composed of either identical or very similar subunits [20, 21]. PAL catalyzes the *trans*-elimination of ammonia and the

Table 6.1 Citrus phenylpropanoid and core flavonoid biosynthetic enzymes

	Description	Accession number	Full length	Enzymology
PAL	*Citrus limon* PAL (pal6) mRNA	U43338	Y [158]	N
	Citrus clementina X *Citrus reticulata* PAL (Fpal1) mRNA	AJ238753	Y [28]	Y [28, 32]
	Citrus clementina X *Citrus reticulata* PAL (Fpal2) mRNA	AJ238754	Y [28]	Y [28, 32]
	Citrus sinensis PAL mRNA	DQ088064	N [101]	N
	Citrus limon PAL mRNA	AY681119	N [160]	N
C4H	*Citrus paradisi* C4H mRNA	AF378333	Y	N
	Citrus sinensis C4H CYP73 (C4H1) mRNA	AF255013	Y [45]	N
	Citrus sinensis C4H CYP73 (C4H2) mRNA	AF255014	Y [45]	N
4CL	–	–	–	–
CHS	*Citrus unshiu* CHS mRNA	FJ887898	Y	N
	Citrus sinensis CHS mRNA (CitCHS2)	AB009351	Y [56]	N
	Citrus sinensis CHS mRNA (CitCHS1)	AB009350	Y [56]	N
	Citrus jambhiri CHS mRNA	AB050890	N [159]	N
CHI	*Citrus unshiu* CHI mRNA	FJ887897	Y	N
	Citrus sinensis CHI mRNA	AB011794	Y [57]	N
	Citrus maxima CHI mRNA	FJ976039	N	N
	Citrus sinensis fruit tissue CHI protein	–	–	Y [66]
	Citrus paradisi fruit tissue chalcone cyclase protein	–	–	Y [64]
F3H	*Citrus maxima* cultivar Guanxi pomelo F3H mRNA	GQ121373	Y	N
	Citrus sinensis CitF3H F3H mRNA	AB011795	Y [57]	Y [73]
	Citrus paradisi leaf tissue F3H protein	–	–	Y [76]
FNS	–	–	–	–
FLS	*Citrus unshiu* CitFLS FLS mRNA	AB011796	Y [92]	Y [93, 95]
DFR	*Citrus sinensis* cultivar tarocco DFR gene	DQ084723	Y [102]	Y [102]
	Citrus sinensis cultivar navel DFR gene	DQ084722	Y [102]	N
	Citrus sinensis DFR mRNA	AY519363	Y [102]	Y [102]
ANS	*Citrus sinensis* ANS mRNA	AY581048	Y	N
	Citrus sinensis ANS mRNA	AY500593	N [109]	N
F3′H	*Citrus sinensis* callus F3′H protein	–	–	Y [113]
F3′5′H	–	–	–	–

pro-3S hydrogen from L-phenylalanine to generate *trans*-cinnamate without the need for external cofactors. The enzyme is inhibited by the product of the reaction, *trans*-cinnamate, suggesting the potential for feedback inhibition [18]. The mechanism of PAL activity is an active area of research that has been greatly aided by solving the crystal structures of PAL from *Rhodosporidium toruloides* (a yeast) and *Petroselinum crispum* (parsley) [22, 23].

In the Rutaceae, PAL activity was initially identified from flavedo disks of *C. paradisi* fruit [24]. The ability of PAL to produce *trans*-cinnamic acid was shown

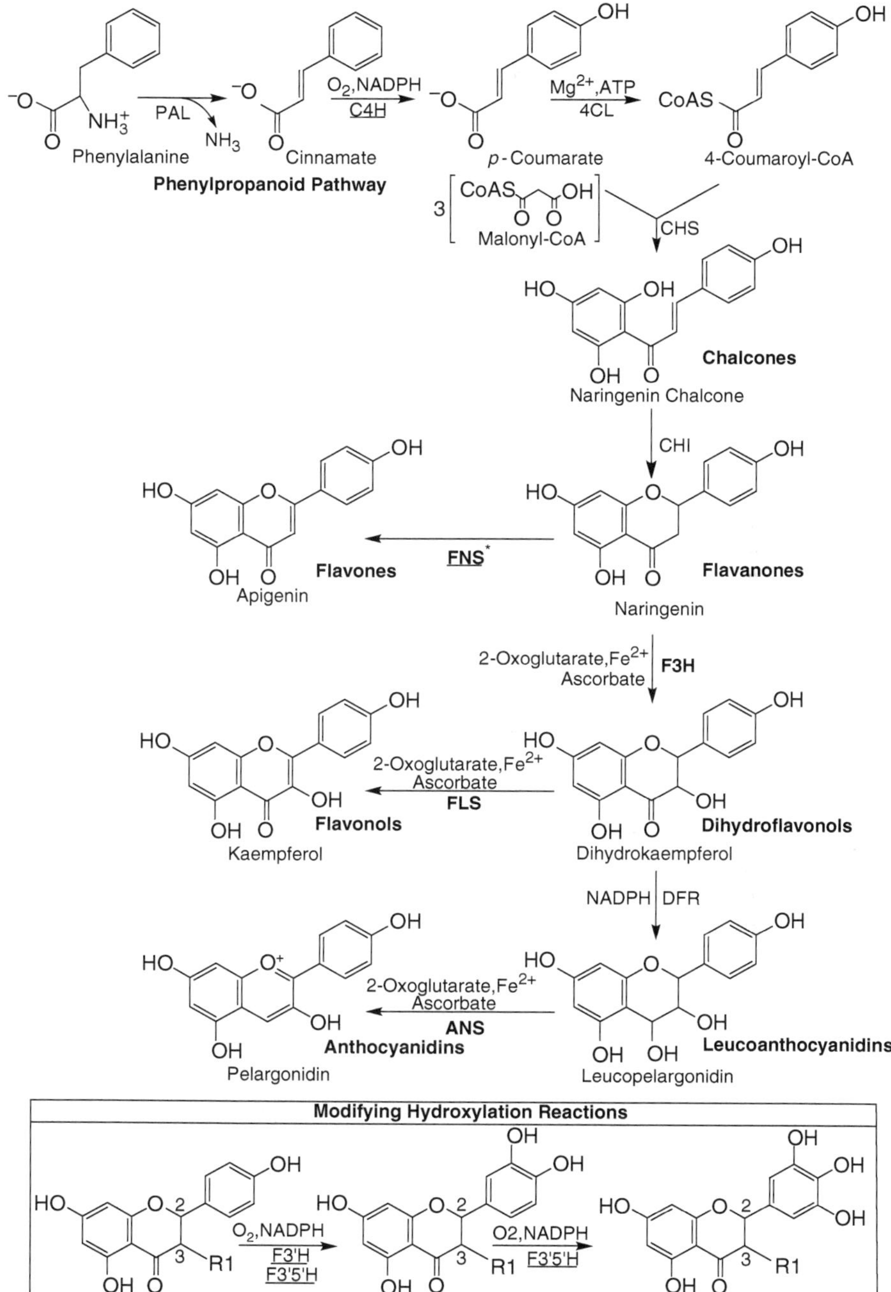

Fig. 6.2 The flavonoid core biosynthetic pathway. The P450s and 2-oxoglutarate-dependent dioxygenases (2-ODDs) are indicated by underlined and bolded titles, respectively. F3′H and F3′5′H are capable of using flavanones (2–3 = single bond, R1 = H), flavones (2–3 = double bond, R1 = H), dihydroflavonols (2–3 = single bond, R1 − OH), or flavonols (2–3 = double bond, R1 = OH) for a substrate. *FNS activity has been indicated with two different enzymes, FNS-II (P450) and the less-common FNS-I (2-ODD)

to decrease as *C. paradisi* fruit size increased and was positively correlated with the accumulation of naringenin glycosides in whole fruit. PAL isolated from acetone powders of *C. paradisi* fruit showed a pH optimum of 9.3 [25]. PAL activity in callus cultures was demonstrated for the first time from grapefruit, and subsequently in *Citrus limon*, Burm (lemon); *C. sinensis*, Osbeck (navel orange); *Citrus reticulata*, Blanco (clementine X Wilking mandarin hybrid); and *Fortunella* sp. X (*C. sinensis* X *Poncirus trifoliata*, Thomasville citrangequat) [26]. Grapefruit PAL activity was not light activated [27], but light activation was observed for PAL in navel orange and Thomasville citrangequat [26].

In *C. clementina* Hort. ex Tanaka x *C. reticulata*, Blanco (fortune mandarin), two PAL genes (AJ238753 and AJ238754) were isolated from a flavedo cDNA library and subsequent Southern blot analysis confirmed that two or more copies of the gene were present in the genome [28]. *In silico* analysis of expressed sequence tags (ESTs) from *C. sinensis* (L) Osbeck identified three putative PAL isoforms in this species [29]. Most species analyzed to date (reviewed in 30, 31) have also shown that PAL is encoded by a multigene family.

In citrus, PAL has been investigated in relation to its possible effects upon fruit storage and quality. For example, ethylene treatment during cool storage temperatures of fruit was shown to increase PAL activity, and this was postulated to serve as a defense mechanism against chilling, thereby increasing storage times [27, 32–34]. It was also demonstrated that γ-irradiation enhanced PAL activity and it was proposed that this would lead to accumulation of phenolic compounds which may, in turn, increase fruit quality upon storage [35].

6.2.2 *Cinnamate 4-Hydroxylase*

Cinnamate 4-hydroxylase (C4H; EC 1.14.13.11, also defined as CYP73A [36]) catalyzes the *p*-hydroxylation of *trans*-cinnamate to form *trans*-p-coumarate. The enzyme has a requirement for molecular oxygen and NADPH as well as association with the electron donor NADPH-cytochrome P450 reductase (CPR; EC 1.6.2.4) for activity. C4H was the first characterized plant P450 [37, 38].

C4H is relatively unstable, present in fairly low abundance, and membrane bound, making it a challenging enzyme for analysis. C4H was initially purified from *Helianthus tuberosus* (Jerusalem artichoke) [39] and *Vigna radiata* (mung bean) [40] using newly developed chromatographic techniques. Subsequently, the first C4H gene sequences were isolated from Jerusalem artichoke [41], mung bean [40], and *Medicago sativa* (alfalfa) [42]. Within the cell, C4H appears to be localized to the endoplasmic reticulum (ER) [43] and has been predicted to serve as one anchor for the proposed phenylpropanoid/flavonoid multienzyme complex [44].

Two full-length cDNAs for C4H, C4H1 (AF255013) and C4H2 (AF255014), have been isolated from Valencia orange, and this was the first time that differential expression patterns were demonstrated for C4H isoforms [45]. C4H2 was constitutively expressed and predicted to have housekeeping functions in the phenylpropanoid pathway, while C4H1 was found only in flavedo and leaf tissue after wounding. C4H2 was expressed at levels 3–10 times higher than C4H1, even after wound induction of C4H1 expression. Southern blot analysis indicated that C4H2 is a single-copy gene, while C4H1 might be represented by two copies in the genome.

6.2.3 *4-Coumarate:CoA Ligase*

The 4-coumarate:CoA ligase (4CL; EC 6.2.1.12) enzyme activates 4-coumaric acid, caffeic acid, ferrulic acid, and (in some cases) sinapic acid by the formation of CoA esters that serve as branch-point metabolites between the phenylpropanoid pathway and the synthesis of secondary metabolites [46, 47]. The reaction has an absolute requirement for Mg^{2+} and ATP as cofactors. Multiple isozymes are present in all plants where it has been studied, some of which have variable substrate specificities consistent with a potential role in controlling accumulation of secondary metabolite end-products. Examination of a navel orange EST database (CitEST) for flavonoid biosynthetic genes resulted in the identification of 10 tentative consensus sequences that potentially represent a multi-enzyme family [29]. Further biochemical characterization will be necessary to establish whether these genes have 4CL activity and, if so, whether preferential substrate usage is observed.

6.2.4 *Chalcone Synthase*

Chalcone synthase (CHS; 2.3.1.74) was the first type III polyketide synthase identified. It is considered to be the key, committing enzyme for flavonoid biosynthesis as it produces the C_{15} backbone, which is derived to form all flavonoid compounds [21]. *In vivo*, CHS catalyzes the condensation of three molecules of malonyl-CoA with 4-coumaroyl-CoA to form naringenin chalcone without the need for cofactors [48]. *In vitro* activity with other hydroxycinnamic-CoA esters often has been observed. CHS activity was initially identified in parsley cell suspension cultures, and the enzyme was purified and a thorough biochemical characterization was conducted [49, 50]. During initial biochemical characterization, CHS was initially misnamed flavanone synthase since the predominant product from the reaction was naringenin. At that time, the spontaneous enantiomeric cyclization of naringenin chalcone to naringenin in solution was not known. The reaction mechanism for CHS has been intensively studied, and a crystal structure of the *Medicago sativa* (alfalfa) CHS homodimer has been solved, allowing further refinement of these studies [51, 52].

The first nucleotide sequence for CHS was determined from cultured parsley cells [53]. Since then, its expression and regulation have been extensively studied in numerous plant systems in relation to a myriad of different conditions and stimuli [e.g., 21 and ref therein]. Due to its key role in flavonoid biosynthesis, CHS has been a popular target for various gene-silencing techniques in attempts to generate flavonoid-deficient plants [16, 54 and ref therein].

In Citrus, CHS activity was first identified from cell-free extracts of calamondin (*C. reticulata* var. *austera* x *Fortunella* sp. = *Citrus mitis* Blanco = *Citrofortunella mitis* = *Citrus madurensis*), and all of the radioactive-labeled products generated from the reaction corresponded to naringenin by TLC analysis [55]. The detection of naringenin instead of naringenin chalcone was likely due to autocyclization as previously described. CHS enzyme activity levels appeared to parallel flavanone glycoside accumulation in young leaves and fruit [55].

In Valencia orange, two CHS clones (CitCHS1; AB009350 and CitCHS2; AB009351) have been identified and differential mRNA expression patterns observed [56]. The expression of CitCHS2 was shown to correlate with the accumulation of flavonoid compounds during embryogenesis, while CitCHS1 expression was barely detectable. Subsequent investigations in Satsuma mandarin (*Citrus unshiu* Marc. cv. Miyagawase) demonstrated expression of the CitCHS1 homolog in young leaves, mature leaves, flowers, and fruitlets while expression of the CitCHS2 homolog was seen in young leaves, flowers, fruitlets, and albedo. This demonstrated varying expression patterns throughout the whole plant and fruit [57]. Recently, the CitCHS2 sequence was used to isolate a CHS gene (EU410483) from *C. sinensis* (L.) Osbeck cv. Ruby (ruby blood orange), and three copies of CHS were shown to exist in the genome by fluorescent *in situ* hybridization (FISH) [58]. AB009351 and EU410483 vary by only 11 nucleotides, which result in only 4 changes at the amino acid level.

In an attempt to influence flavor characteristics, CHS antisense and overexpression constructs were introduced into grapefruit by *Agrobacterium tumefaciens* [16, 59]. CHS overexpression constructs were shown to lead to morphological abnormalities and frequently to death, while a number of CHS antisense constructs resulted in a statistically significant reduction in flavonoid content in the transgenic plants even though there was a large variation in the flavonoid concentration [16].

6.2.5 Chalcone Isomerase

The stereospecific cyclization of chalcones to (2S)-flavanones is a prerequisite for the synthesis of the majority of flavonoid subclasses derived from this branch point metabolite. This reaction is catalyzed by chalcone isomerase (CHI, CFI; EC 5.5.1.6). CHI exists in two forms, one that accepts only 6′-hydroxychalcones and another that accepts both 6′-hydroxy-(naringenin chalcone) and 6′-deoxychalcones (isoliquirgentin), the latter generally found in legumes. Although 6′-hydroxychalcones will spontaneously convert to a racemic flavanone mixture, the CHI-catalyzed reaction proceeds at a rate 36 million-fold faster and is highly stereoselective for the formation of (2S)-flavanones [60]. Spontaneous isomerization of 6′-deoxychalcones does not substantially occur without enzyme catalysis.

The enzyme was initially purified from soybean (*Glycine max* = *Soja hispida*) and biochemically characterized primarily using isoliquirgentin as the substrate [60, 61]. CHI was active as a monomer and no cofactors were required for enzyme activity. A crystal structure for alfalfa CHI in complex with naringenin has been solved [62].

CHI activity was first identified in lemon (*C. limon*) peel extracts with crude chalcone preparations chemically derived from the flavonoid glycosides naringin (isolated from *Citrus natsudaidai*) and poncirin (isolated from *Poncirus trifoliata*), as well as the aglycone, isosakuranin (synthesized from poncirin) [63]. Extracts from acetone powders of *C. paradisi* fruit were shown to convert isoliquiritigenin (6′-deoxychalcone) to liquiritigenin [25]. A number of other unidentified spots were noted during TLC analysis of these CHI reaction products. The activity with isoliquiritigenin could not be verified in subsequent experiments with *C. paradisi* fruit tissues; however, chalcone cyclase (CC) activity was noted for chalcone glycosides bearing a 4′-neohesperidoside [64].

As all other CHI enzymes characterized only accept aglycones as substrates, the discovery of CC activity led to the hypothesis that an alternate form of the flavonoid biosynthetic pathway, in which glycosylation occurs before cyclization of the chalcone, may exist in Citrus. Resolution was further complicated when it was demonstrated that chalcones did not need to be glycosylated before their conversion to flavanones in calamondin [55]. It was then shown that flavanone UDP-glucosyltransferase and UDP-rhamnosyltranferase activities exist in cell free extracts of calamondin, pummelo, also called shaddock (*Citrus maxima* = *Citrus grandis*), and grapefruit, demonstrating that glycosylation could occur after the cyclization reaction [55, 65]. More recently, Fouche and Dubery [66] have purified (to apparent homogeneity) and biochemically characterized a CHI from *C. sinensis* that only accepts naringenin chalcone as a substrate. Thus, most current evidence supports the existence of a citrus flavonoid biosynthetic pathway that parallels that which has been observed in other plants. The significance of the CC activity still needs to be resolved.

A cDNA clone for CHI was isolated from young Valencia orange seed and used to generate probes for Southern and northern analyses [57]. Southern blot analysis suggested that there was only one copy of CHI in the satsuma mandarin genome. Northern analysis demonstrated high expression levels in young developing tissue, which decreased in senescing tissue and during fruit development. These patterns were paralleled by the CitCHS1, CitCHS2, and CitF3H expression seen in Valencia orange [57]. CHI-sense and CHI-antisense constructs have been introduced into grapefruit in an attempt to influence fruit flavor characteristics with varying levels of success. CHI-antisense constructs had no significant effect upon flavonoid accumulation in leaf tissue, while CHI-sense constructs did have a statistically significant impact upon flavonoid accumulation, likely due to co-suppression effects [59, more detailed information in 16].

6.2.6 Flavanone 3β-Hydroxylase

Flavanone 3β-hydroxylase (F3H, FHT; EC 1.14.11.9) is a selective enzyme that catalyzes the stereospecific 3β-hydroxylation of 2S-flavanones to form 2R,3R dihydroflavonols [21]. The enzyme requires 2-oxoglutarate, Fe^{2+}, and ascorbate as cofactors and was the first 2-oxoglutarate-dependent dioxygenase identified in the flavonoid pathway. F3H was initially characterized in parsley (*Petroselinum crispum*) and *Matthiola incana* [67]. The majority of the subsequent biochemical characterization was carried out with the *Petunia hybrida* protein [68]. F3H is active as a monomeric protein, and several studies have identified amino acid residues that are important for activity [69–72]. This work has been hampered by the fact that the enzyme undergoes proteolytic degradation in plant extracts as well as in recombinant expression systems [69–73]. In addition to Petunia F3H, tea (*Camellia sinensis*), apple (*Malus domestica* cv. 'M9'), and *Arabidopsis thaliana* F3H enzymes have been investigated thoroughly at the biochemical level [73–75].

Although citrus F3Hs have not been biochemically characterized in depth, conversion of naringenin to dihydrokaempferol has been established for a recombinantly expressed orange F3H [73] and from crude extracts of young grapefruit leaves [76]. The latter analysis focused predominantly on developing an assay for the enzyme employing capillary electrophoresis.

A full-length F3H cDNA clone was isolated from seeds of Valencia orange and used to generate probes to analyze F3H in satsuma mandarin [57]. Southern blot analysis suggested that F3H may exist in multiple copies in the mandarin genome. Northern blot analysis indicated expression levels were highest in young, rapidly growing tissues such as young leaves and fruitlets. In grapefruit, F3H was demonstrated to be a single copy gene, and expression across multiple developmental stages and tissues was examined by RT-PCR [77]. Petunia F3H expression levels were examined in parallel, and it was shown that grapefruit F3H was expressed at higher levels in all tissues examined. This suggested that accumulation of flavanone glycosides in grapefruit, which have not been identified in petunia, was not due to lowered F3H expression levels and was more likely due to biochemical properties of grapefruit flavonoid glycosyltransferases [77].

6.2.7 Flavone Synthase

Flavone synthase (FNS; EC 1.14.11.22) introduces a double bond between C2 and C3 of a flavanone to produce the corresponding flavone. This activity was initially identified in parsley cell suspension cultures and subsequently shown to be encoded by a 2-oxoglutarate-dependent dioxygenase [67, 78, 79]. This enzyme, now known as FNS-I, appears to have very limited distribution. To date, it has only been identified in the Apiaceae family (Umbellifers). The more widely occurring FNS-II (CYP93B) was initially identified from snapdragon (*Antirrhinum majus*) flowers [80] and was subsequently shown to be a P450 enzyme. FNS-I, FNS-II, and the various roles flavones play in plant species have recently been reviewed by Martens and Mithöfer [81]. Subsequent to this review, Yu et al. [82] demonstrated that the characteristic lack of natural accumulation of flavones in Brassicaceae could not be overcome in *A. thaliana* even by overexpression of recombinant parsley FNS-I.

However, exogenous naringenin supplied in the growth media did lead to significant accumulation of apigenin within the plants. It remains to be seen if similar results would be observed if FNS-II was used in a comparable set of experiments.

Citrus species are well-known for their accumulation of flavone- and flavanone-glycosides, and thus should contain all of the enzyme activities necessary for the synthesis of these compounds. Two tentative consensus sequences for FNS-II have been identified by *in silico* analysis of the CitEST database, apparently representing the first identification of putative FNS-II genes in this genus [29]. Biochemical determination of function and analysis of the proteins encoded by these genes will be an important step toward elucidating flavone synthesis in Citrus.

6.2.8 Flavonol Synthase

Flavonol synthase (FLS; E.C.1.14.11.23) catalyzes the committed step in the production of flavonols by introduction of a double bond between C2 and C3 of the corresponding dihydroflavonols. Like F3H, FLS has been described as a 2-oxoglutatarate-dependent dioxygenase based on its cofactor requirements for 2-oxoglutarate, Fe^{2+}, and ascorbate. FLS was initially identified in enzyme preparations from illuminated parsley cell suspension cultures [67]. Subsequently, FLS was characterized from the flower buds of *Matthiola incana* and carnation (*Dianthus caryophyllus* L.), and it was suggested that there was regulation between flavonol and anthocyanidin biosynthesis [83, 84].

The first *FLS* gene identified was isolated by differential screening of a *P. hybrida* cDNA library, and gene identity was confirmed by performing enzyme activity assays with recombinant protein [85]. In *Arabidopsis thaliana*, FLS is the only central flavonoid biosynthetic pathway enzyme to be encoded by multiple isoforms [86]. Of the six FLS genes identified in the Arabidopsis genome, fls2, fl4, and fls6 are thought to encode pseudogenes as they lack residues that were previously shown to be required for biochemical activity in dioxygenases [87–89]. Only one of the predicted functional isoforms (FLS1) was actually shown to have enzymatic activity with *in vitro* assays, and only the corresponding FLS1 T-DNA knockout plant lines were shown to impact flavonol accumulation *in planta* [87, 90]. Although FLS3 has no measurable activity in standard *in vitro* assays and although T-DNA knockout lines have no significant impact on *in planta* flavonol accumulation, it has been shown that recombinant yeast cells expressing FLS3 cultured with dihydroflavonols in the media (20 mg substrate per 50 mL of culture) will produce detectable quantities of flavonols after highly extended (48 h) incubation times [91].

Much of the FLS biochemical and structure/function analysis has focused on a protein from *C. unshiu* (mandarin). A cDNA was isolated based on sequence homology to an Arabidopsis *FLS* EST (153O10T7) and used as a probe to determine regulation of gene expression [92]. Higher expression was observed in young leaf, flower, peel, and juice sac/segment epidermis tissues. Expression decreased with tissue age, as has been observed for citrus CHS, CHI, and F3H [57]. Southern analysis using a mandarin FLS-specific probe suggested that several FLS genes or related sequences may be present in this species. A thorough biochemical characterization of recombinant mandarin FLS protein showed that it had a sixfold higher affinity for dihydrokaempferol than dihydroquercetin [93]. Site-directed mutagenesis revealed two FLS residues that are strictly conserved across species, Gly68 and Gly261, were important for enzyme activity and were likely involved in proper folding of the enzyme. The *C. unshiu* enzyme has also been shown to have promiscuous activity with naringenin, as previously identified for Arabidopsis FLS [94, 95]. This promiscuous acceptance of substrates is in sharp contrast to the strict regiospecificity of the early pathway dioxygenase, F3H.

6.2.9　Dihydroflavonol 4-Reductase

Dihydroflavonol 4-reductase (DFR; EC 1.1.1.219) is a member of the short-chain dehydrogenase/reductase family and catalyzes the stereospecific conversion of (+)-(2R,3R)-dihydroflavonols to the corresponding (2R,3S,4S) flavan-3,4-*cis*-diols (leucoanthocyanidins), with NADPH as a required cofactor. The enzyme activity was first identified in cell suspension cultures of Douglas fir (*Pseudotsuga menziesii*) and was shown to be related to the accumulation of flavan-3-ols and proanthocyanidins [96]. Leucoanthocyanidins and DFR were later shown to be required for anthocyanidin formation by complementation of *Matthiola incana* mutants blocked between dihydroflavonol and anthocyanidin biosynthesis [97, 98]. DFR has been purified to apparent homogeneity and biochemically analyzed from flower buds of *Dahlia variabilis* [99]. DFR was shown to accept different substrates depending on the plant species from which it was isolated (reviewed in 100).

Although Citrus species do not generally produce anthocyanins, the blood orange cultivars (e.g. Tarocco, Moro, and Sanguinello) are an exception, being defined by their accumulation of anthocyanins in the fruit flesh and peel. DFR, CHS, and a predicted anthocyanidin 3-O-GT were shown by RT-PCR to be up-regulated in response to cold temperature exposure in Tarocco [101]. In a following study, genomic and cDNA clones for DFR were isolated from blood orange (Tarocco) and blond orange (Navel and Ovale) fruit, and Southern blot analysis showed that the gene is present as a single copy in all samples analyzed [102]. The DFR sequences of blood and blond orange were shown to be 100% identical, and enzyme activity was demonstrated for a DFR clone expressed recombinantly in *Escherichia coli*. [102]. Therefore, the lack of accumulation of anthocyanins in blond oranges is not due to an inactive DFR enzyme. However, expression of DFR as quantified by RT-PCR was approximately 250-fold lower in blond oranges as opposed to blood cultivars. Additionally, DFR enzyme activity was measurable only in crude lysates from blood orange fruit. This has led to the hypothesis that the variation in anthocyanin production between blond and blood oranges may result from a mutation in a regulatory gene that impacts DFR expression. Because no Citrus has been shown to produce proanthocyanidins or catechins (alternative fates for leucoanthocyanidins), the potential function of a gene encoding an active DFR enzyme in blond orange remains unknown.

6.2.10 Anthocyanidin Synthase

Anthocyanidin synthase (ANS, LDOX; EC 1.14.11.19) is a 2-oxoglutarate-dependent dioxygenase that catalyzes the conversion of leucoanthocyanidins to the colored anthocyanidins. ANS was initially identified by its ability to complement the A2 mutation in corn (*Zea mays*), which was known to be blocked in the ability to convert leucoanthocyanidins to anthocyanidins [103, 104]. ANS was initially classified as a dioxygenase based on sequence similarity, and this was subsequently confirmed in *Perilla frutescens* and Arabidopsis by direct biochemical evidence (including cofactor requirements of 2-oxoglutarate, Fe^{2+}, and ascorbate) using recombinant enzymes expressed in *E. coli* [105, 106]. ANS has been proposed to have promiscuous substrate usage, and *in vitro* experiments with recombinant enzyme using *cis-* and *trans*-dihydroquercetin predominantly produced flavonols with anthocyanidins as a minor product [107, 108].

In Citrus, *in silico* analysis suggested one copy of the ANS gene in orange [29]. In a variety of blood orange cultivars, mRNA accumulation for ANS as well as for CHS and a proposed anthocyanidin 3-O-GT was shown to be positively correlated with anthocyanidin accumulation [109]. It was further suggested that these genes are not missing from blond orange, but downregulated at the level of mRNA expression. Clearly, much remains to be discovered for ANS in Citrus.

6.2.11 Flavanone 3′-Hydroxylase

Flavanone 3′-hydroxylase (F3′H; EC1.14.13.21; CYP75B) activity was initially identified in microsomal preparations of golden weed (*Haplopappus gracilis*) [110]. F3′H from irradiated parsley cell cultures was later biochemically analyzed and characterized as a cytochrome P450 having an absolute requirement for NADPH and molecular oxygen as cofactors [111]. The enzyme has been shown to have activity with flavanones, flavones, dihydroflavonols, and flavonols, but does not appear to have activity with anthocyanidins [111]. The first cDNA clone for F3′H was isolated from Petunia [112]. It has been suggested that F3′H may serve as an anchor for the proposed flavonoid multi-enzyme complex on the cytosolic surface of the endoplasmic reticulum [44].

F3′H activity from the microsomal fraction of cell cultures of *C. sinensis* (L.) Osbeck cv. Hamlin (Hamlin) has been biochemically characterized and verified to be a cytochrome P450 [113]. The enzyme was shown to use naringenin, dihydrokaempferol, and kaempferol, but not apigenin as substrates. F3′H activity was also demonstrated from young leaves of Hamlin orange as well as from the flavedo of Hamlin orange, Marsh grapefruit, and Lisbon lemon fruits [113]. The fact that F3′H activity was present in young leaves but not in older Hamlin leaves is consistent with what has been seen previously for other flavonoid pathway enzymes.

6.2.12 Flavanone 3′,5′-Hydroxylase

Flavanone 3′,5′-hydroxylase (F3′5′H; EC 1.14.13.88; CYP75B) activity was first reported from cell-free extracts of *Verbena hybrida* flowers [114] and, like F3′H, it has been characterized as cytochrome P450 based on its cofactor requirements (NADPH, O_2) and inactivation by P450 inhibitors [48]. The enzyme can hydroxylate naringenin and dihydrokaempferol (4′-OH-containing compounds) at the 3′ and 5′ positions as well as eriodictyol and dihydroquercetin (3′,4′-OH containing compounds) at the 5′ position to form the corresponding 3′,4′,5,5′,7-pentahydroxyflavanone and dihydromyricetin. F3′5′H is not ubiquitous, but it has been identified in a number of ornamental species and has been studied extensively in relation to its role in generating blue flowers [100].

There have been no apparent reports of F3′5′H from any citrus species in the scientific literature or gene sequences deposited in available databases. Also, no strong candidate consensus sequence has been identified in the citrus HarvEST databases after conducting homology searches with various known F3′5′H gene sequences from non-citrus species (Owens, unpublished). Although the 2003 USDA Database for the Flavonoid Content of Selected Food indicates that myricetin, which requires F3′5′H activity for its synthesis, has been identified at low levels in orange, grapefruit, and lemon juices, a closer review of the primary literature revealed that the concentrations were actually below the level of detection for the RP-HPLC technique used [115]. However, the identification of delphinidin 3-glucoside, delphinidin 3,5-diglucoside, delphinidin 2-(6′-malonylglucoside), petunidin 3-glucoside, and petunidin-3,5 glucoside as minor pigments in the juices of blood orange varieties suggests that F3′5′H activity is likely present in blood oranges [116, 117, 118, 119].

6.3 Flavonoid Glycosylation in Citrus

As discussed previously, the vast majority of naturally occurring flavonoids in citrus are present in glycosylated forms, with the most abundant being flavanone- and flavone-7-O-glycosides. C-glycosylated flavonoids are also present [2, 120] but in lower abundance. Early studies on flavonoid glycosylation focused on results from radiotracer incorporation or on demonstration of glycosyltransferase activity. For example, Lewinsohn et al. [121, 122] demonstrated 7-O-glucosylation of exogenous naringenin by citrus liquid suspension cultures (Table 6.2). This group also demonstrated 7-O-glucosylation *in vitro* using crude cell extracts [55].

McIntosh and Mansell [65] identified and partially purified flavanone-7-, flavone-7-, chalcone-4′-, and flavonol-O glucosyltransferases from young *C. paradisi* leaves.

Table 6.2 Biochemical properties of Citrus flavanone-7-O-Glucosyltransferases

Property	*Citrus paradisi* cell culture[a]	*Citrus mitis* and *Citrus maxima* young fruit/leaves	*Citrus paradisi* young leaves[b]	*Citrus limon* several tissues
Purification factor	NA	NA	1,476	Not reported
Primary substrates (K_m^{app}, μM)	naringenin hesperetin	naringenin hesperetin	naringenin (62) hesperetin (124)	hesperetin apigenin, crysin kaempferol, morin
Other compounds tested			luteolin, apigenin kaempferol, quercetin rhamnetin, naringenin chalcone	
Sugar donor (Km^{app}, μM)			UDP-glucose (51 with nae 124 with hae)	UDP-glucose (14 with hae)
pH optimum			7.5–8.0	
pI^{app}			4.3	
MW (kDa)			54.9	
Inhibitors (Ki^{app})			UDP (>10 mM) UTP, ATP, TTP	
Ref.	[121, 122]	[55]	[65, 123]	[124]

[a]Biotransformation study.
[b]Flavanone specific.

They subsequently purified a flavanone-specific 7-O-glucosyltransferase to near homogeneity (Table 6.2), resolved it from other glucosyltransferase enzymes in grapefruit seedlings, and characterized its biochemical and kinetic properties [123]. They also reported properties of other flavonoid glucosyltransferases in grapefruit seedlings. The flavanone-specific 7-O-glucosyltransferasee is responsible for catalyzing the first sugar transfer reaction in the synthesis of flavanone-7-O-neohesperidosides or rutinosides [65, 123] in grapefruit. Berhow and Smolensky [124] demonstrated the existence of an analogous enzyme in both *C. limon* leaves and flowers, although levels of activity were 30-fold higher in young leaf tissue (Table 6.2).

Flavanone-specific 7-O-glucosyltransferase from grapefruit leaves is a very active yet low abundance enzyme that is over 10,000 times less sensitive to UDP inhibition as compared to other flavonoid enzymes using UDP-glucose as the sugar donor [65, 125, 126, and ref therein]. Because this enzyme is involved in the first aglycone-capturing glycosylation reaction leading to the formation of naringin and other flavanone 7-O-diglycosides in grapefruit, it is logical to postulate that the low sensitivity to inhibition by UDP may make a significant contribution to naringin biosynthesis *in vivo*. At this time it is not known if the analogous enzyme in orange (leading to production of hesperetin 7-O-diglycosides) shares this property. The effect of UDP on reaction kinetics was not determined for the partially purified enzyme from lemon [124]. Because of the potential impact of biochemical regulation of enzymes in the overall regulation of flavonoid metabolism, it is unfortunate that the information on UDP inhibition is not more widely available in the literature. The labile nature of flavonoid glucosyltransferase activity during isolation and purification has no doubt been a confounding factor.

As stated previously, the majority of the flavonoid glycosides found in citrus are flavanone or flavone diglycosides. Isolation of glucosyltransferases in Citrus supports the hypothesis that addition of sugars to form diglycosides is a stepwise process involving separate sugar-transferring enzymes. This was first demonstrated by Jourdan and Mansell [127] when they established that three different glucosyltransferase enzymes were necessary for the synthesis of kaempferol-3-O-triglucoside in *Pisum sativum*. Kleinehollenhorst et al. [128] subsequently established a similar pattern in *Tulipa*. Since then, many other instances have been documented [21 and ref therein].

Because of the economic significance of naringin in grapefruit and hesperidin in orange and lemon, the rhamnosylation reactions involved in the last biosynthetic step leading to synthesis of these compounds are of significant interest (Table 6.3). Lewinsohn et al. [122] showed that crude extracts from *C. maxima* and *C. mitis* could synthesize rhamnoglucoside in addition to 7-O-glucoside in assays using hesperetin and UGP-^{14}C-glucose as substrates. This suggested the presence of an enzyme converting UDP-glucose to UDP-rhamnose was present in these tissues as well as a rhamnosyltransferase. McIntosh and Mansell [65] demonstrated this

Table 6.3 Biochemical properties of Citrus flavanone 7-O-Glucoside, 2″-O-Rhamnosyltransferases

	Citrus paradisi young leaves	*Citrus mitis and Citrus maxima* young fruit/leaves	*Citrus maxima* leaves
Purification factor	NA	NA	2,735
Primary substrates (K_m^{app}, μM)	prunin (via naringenin)	prunin hesperetin-7-glc	prunin (2.4) hesperetin-7-glc (42.5) apigenin-7-glc luteolin-7-glc
Other compounds tested			naringenin hesperetin quercetin naringenin-5-glc
Sugar donor (K_m^{app}, μM)			UDP-rhamnose (1.1–1.3)
pH optimum			ND
pI^{app}			ND
MW (kDa)			52
Inhibitors (K_i^{app}, μM)			UDP (10 μM) UTP hesperetin naringenin quercetin
Gene accession no.			AY048882 [130]
Reference	[65]	[55]	[129, 130]

same phenomenon in *C. paradisi* leaf crude extracts, and the rhamnosyltransferase activity was separated from the glucosyltransferase activity during purification, further supporting the existence of separate enzymes for the glucosylation and rhamnosylation reactions (Table 6.3).

Historically, the study of flavonoid UDP-rhamnosyltransferase (RT) enzymes has been challenged by the lack of a commercial source of UDP-rhamnose. In 1991, Bar-Peled et al. [129] reported on the purification and characterization of UDP-rhamnose:flavanone-7-*O*-glucoside-2″-*O*-rhamnosyltransferase from pummelo leaves (Table 6.3). They synthesized UDP-^{14}C-rhamnose from UDP-^{14}C-glucose in a method that took advantage of the endogenous process in pummelo [129]. This RT enzyme was sensitive to inhibition by UDP (unlike the flavanone-specific glucosyltransferase from grapefruit [65]), catalyzed the 1–2 addition of rhamnose to glucose, and was specific for flavanone- and flavone-7-*O*-glucosides; it did not rhamnosylate naringenin-5-*O*-glucoside (Table 6.3). Activity with other flavonoid-7-*O*-glucosides was not determined. In a more recent study, Frydman et al. [130] isolated the 1-2RT gene (Cm1,2RhaT) from pummelo and established its biochemical function using a transgenic approach (Table 6.3). They demonstrated that cell cultures of the transgenic tobacco line could produce naringin when fed naringenin-7-*O*-glucoside. Further, because the tobacco cell line contained an endogenous broad specificity flavonoid-7-*O*-glucosyltransferase, transgenic cultures fed naringenin also were capable of synthesizing naringin. They reported that while cultures also produced kaempferol- and quercetin-7-*O*-glucosides, neohesperidosides of these compounds were not synthesized by the transgenic line. This supports the specificity of this RT for flavanone- and flavone-7-*O*-glucosides.

One of the intriguing biochemical questions that remain to be solved is that of the structure/function relationships of the 1–2RT and the 1–6RT in Citrus. As an example, consider the rhamnosylation of naringenin-7-*O*-glucoside into either narirutin or naringin (Fig. 6.1). In both instances, the substrates are naringenin-7-*O*-glucoside and UDP-rhamnose. Therefore, substrate recognition sites will be similar, yet one enzyme will catalyze addition of rhamnose to the #2 carbon of glucose to make the bitter linkage and the other enzyme will catalyze addition of rhamnose to the #6 carbon of glucose to make the tasteless diglycosides. RTs catalyzing 1–6 additions have been found in other species but the homology to the 1–2RT from pummelo is low [130]. It is common to see low homology for flavonoid glycosyltransferase enzymes from different species even those catalyzing the same reaction, although those from varietals or related species have higher homology [126, 131, 132, and ref therein]. Frydman and Eyal have posted a cDNA sequence defined as a putative 1–6RT gene from *C. sinensis* (DQ119035); however, biochemical evidence for the function of the protein has not yet been reported. Once biochemical identity is established, comparison of the structures of the 1–2RT and the 1–6RT enzymes and conducting analysis of structural elements conferring these different specificities would provide opportunity for analysis at a deep level and address questions regarding structural characters imparting the specific additions to the #2 or #6 carbon of glucose. It may be especially interesting to compare the RTs in grapefruit or in other species that produce both rutinosides and neohesperidosides.

With the exception of the 1-2RT in pummelo, the approach of obtaining structural information or genes using amino acid sequence information from purified citrus glycosyltransferases has not been successful [133]. The contributing factors are relative low abundance in tissue as well as protein instability through several sequential purification processes. While the Plant Secondary Product Glycosyltransferase (PSPG) box can be used as a sequence marker for identification of putative GTs (Fig. 6.3), the current futility of a purely *in silico* approach for assigning plant GT function based solely on primary sequence information is well known [e.g., 126, 132, 134 and ref therein]. While this has led to efforts to use alternate approaches for *in silico* analysis [e.g., 132, 135], these approaches have not yet realized their full potential. Therefore, alternate approaches for structure/function analysis must be employed.

Recently, we have used several approaches to obtain putative flavonoid GT clones from young grapefruit leaf cDNA. Genes are cloned into pCD1 expression vector [73], transformed into expression cells, and induced to express recombinant protein. The pCD1 vector is a modified pET32 vector that includes a thrombin cleavage site immediately adjacent to the GT, thereby allowing for removal of tags. Thus, recombinant protein and protein with the tags removed can be assessed for GT activity.

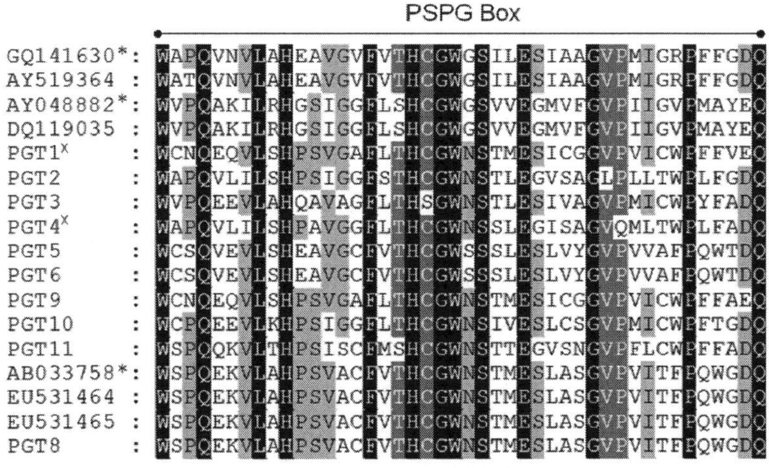

Fig. 6.3 Alignment of PSPG box sequences from Citrus flavonoid glycosyltransferases and other putative secondary product glucosyltransferase clones. The aligned regions are from *Citrus paradisi* cv. Duncan flavonol 3-O-GT (GQ141630), *Citrus sinensis* putative flavonoid 3-O-GT (AY519364), *Citrus maxima* flavonoid 1–2 RT (AY048882), *Citrus sinensis* putative 1–6 RT (DQ119035), *Citrus paradisi* cv. Duncan putative glucosyltransferases (PGT1-6,9-11) *Citrus unshiu* limonoid GT (AB033758), *Citrus paradisi* cv. Marsh putative limonoid GT (EU531464), *Citrus sinensis* cv. Shahsavari putative limonoid GT (EU531465), and *Citrus paradisi* cv. Duncan putative limonoid GT(PGT8). Astericks (*) indicate that function was established biochemically and 'x' denotes current evidence indicates GT does not act on flavonoid substrates

Of the 11 clones we currently have in-hand (Fig. 6.3), 1 has been shown to encode a protein with flavonol-specific 3-*O*-glucosyltransferase activity [126] and its properties are summarized in Table 6.4. Two clones encode proteins that did not have activity with flavonoid substrates [131; McIntosh, unpublished results]. Another clone has over 95% identity to limonoid-17-*O*-glucosyltransferases in *C. unshiu* [136] and the activity of the grapefruit homolog is being determined. We have full-length clones for five other putative flavonoid GTs, and these are in different stages of characterization at this time. Determination of final consensus sequences for the remaining two clones is in progress. In spite of using the pCD1 vector containing a thioredoxin tag to increase solubility of protein expressed in an *E. coli* system, there is a consistent issue with the majority of expressed protein being packed into bacterial inclusion bodies. This is not an unusual occurrence for secondary product GTs [126, 137, Mallampalli, unpublished; Blount, personal communication]. For some clones, sufficient soluble protein for analysis of enzymatic activity can be obtained through manipulation of culture conditions. For others, we are exploring alternate expression systems.

The importance of rigorous testing of enzymatic activity of the recombinant proteins cannot be overemphasized. To fully evaluate the binding ability, catalytic specificity, and reaction kinetics and to use information obtained from *in vitro* analyses to inform assignment of *in vivo* function, it is important to test many flavonoid classes as potential substrates and conduct thorough identification of reaction product(s). Such an approach was used to assign function of the flavonol-specific-3-*O*-glucosyltransferase clone from grapefruit [126]. Testing clone function through direct assay of the protein is a critical aspect in order to correlate results

Table 6.4 Biochemical properties of Citrus flavonol 3-*O*-glucosyltransferases

	Citrus paradisi, leaf heterologous expression	*Citrus sinensis* (Tarocco)[a] fruit
Purification Factor	NA	
Primary substrates (K_m^{app}, μM)	kaempferol (12); quercetin (67); myricetin (33)	
Other compounds tested	cyanidin, apigenin, luteolin 4′-methoxyflavonol 3′,4′-dimethoxyflavonol 4′acetoxy-7-OH-6-methoxyisoflavone naringenin dihydroquercetin	
Sugar donor (K_m^{app}, μM)	UDP-glucose (669)	
pH optimum	7.5	
pI^{app}	6.27	
MW (kDa)	51.2	
Inhibitors (K_i^{app})	UDP (69.5 μM) Cu^{2+}, Fe^{2+} Zn^{2+}	
Gene accession no.	GQ141630 [126]	AY519364 [101]
Reference	[126]	

[a]Biochemical characterization of recombinant protein in progress (Owens et al., unpublished).

from biochemical analyses of enzymes isolated from plant tissues with results from putative GTs identified through molecular and/or bioinformatic techniques as was done for the 1–2RT [129, 130].

6.3.1 Specificity of Flavonoid Glycosyltransferases in Citrus

There have been reports of promiscuous flavonoid GTs in plants as well as reports of highly specific enzymes [21 and ref. therein, 123, 125, 132, 137–142]. In some cases, apparent promiscuity could be the result of the presence of other contaminating GTs in the preparation. Presence of a large tag in recombinant enzymes also raises a question about the relationship of observable characters of the tagged recombinant enzyme to the role of the native enzyme *in vivo*. It is notable, however, that some rigorously purified and characterized enzymes have also exhibited promiscuous use of flavonoid substrates. A report of expression of a promiscuous flavonoid GT induced by treatment with salicylic acid may indicate multiple strategies for expression and activity of different GTs depending on life stage and stress levels [143]. To date, citrus flavonoid GTs that have been purified to a significant level of enrichment or apparent homogeneity, as well as recombinantly expressed and purified GTs, have been shown to be regiospecific and specific to one or at most two flavonoid classes [65, 123, 126, 129, 130]. If these properties exist for other citrus flavonoid GTs, then, considering the variety of flavonoid glycosides in Citrus, there are likely dozens more flavonoid GTs to be found and characterized.

6.4 Function of Citrus Flavonoid Glycosides

The impact of flavanone neohesperidosides on flavor of fruits and processed products of some varieties, the role of flavanone rutinosides in formation of "cloud" in other varieties, and the impact of these compounds on consumer acceptance were discussed above. It is logical to hypothesize that the high level of bitter flavonoids in young leaf and very young fruit tissues may have a role in defense from herbivory, at least until seeds within the fruit are mature, although actual experiments have not been conducted. Recent reviews of general roles of flavonoids in plants include many other possibilities [144, 145]. Some direct testing of roles of flavonoids in citrus have shown polymethoxyflavonoids are key for defense against microbes [146–148 and ref therein].

Because of human consumption of plant and plant products, there has been much interest on the impact of flavonoids on human health and this has been recently reviewed [149, 150 and ref therein]. Additionally, Passamonti et al. [151] have recently reviewed the issue of bioavailability of dietary flavonoids. Citrus flavonoids have been shown to have many beneficial effects on human health including anti-inflammatory activity, anticancer activity, antioxidant activity, and protection against coronary heart disease [reviewed in 152–154]. This has led to renewed efforts to increase efficiency of recovery of bioactive flavonoids from processed products [155, 156]. There is no doubt that additional roles will continue to be discovered in the future as research in this area continues. Citrus contains many other health-promoting compounds in addition to flavonoids and sustained interest in crop improvement is one of the motivating factors for an international Citrus genomics effort. This has been recently reviewed by Talon and Gmitter [157].

6.5 Concluding Remarks and Future Perspectives

Citrus contains many different glycosylated flavonoids with a variety of functions in plants and in human health. The diversity of flavonoid GTs present in Citrus makes it an attractive model system for in-depth structure/function analysis of this group of enzymes. The economic impact of the glycosylated flavonoids, and the potential benefit of up-regulating some and down-regulating others, is a logical driver for exploring biotechnological applications of modifying flavonoid metabolism in Citrus. The challenge is working efficiently with transgenic citrus and to get a handle on the inherent natural variability in synthesis and accumulation of economically important compounds. Because most compounds are in glycosylated form, the impact of GT enzymes must be recognized as an important factor for understanding flavonoid metabolism and the potential impact of perturbing the system.

Acknowledgments The authors gratefully acknowledge the USDA (grants 98-35301-6514 and 2003-35318-13749) and NSF (grant MCB-0614260) for past and current support of their work on secondary product glycosyltransferases and flavonoid metabolism in *C. paradisi*.

References

1. Horowitz RM (1986) Taste effects of flavonoids. In: Cody V, Middleton E, Harborne JB (eds) Plant flavonoids in biology and medicine. Alan R. Liss Inc, NewYork, NY, pp 163–176
2. Berhow M, Tisserat B, Kanes K, Vandercook C (1998) Survey of phenolic compounds produced in Citrus. USDA ARS Technical Bull #158
3. Scora RW, Kumamoto J (1983) Chemotaxonomy of the genus *Citrus*. In: Waterman PG Grundon MF (eds) Chemistry and chemical taxonomy of the rutales. Academic Press, London, pp 343–351
4. Peterson JJ, Dwyer JT, Beecher GR, Bhagwat SA, Gebhardt SE, Haytowitz DB, Holden JM (2006) Flavanones in oranges, tangerines (mandarins), tangors, and tangelos: a compilation and review of the data from the analytical literature. J Food Compost Anal 19:S66–S73
5. Gattuso G, Barreca D, Gagiulli C, Leuzzi U, Caristi C (2007) Flavonoid composition of *Citrus* juices. Molecules 12:1641–1673
6. Djoukeng JD, Arbona V, Argamasilla R, Gomez-Cadenas A (2008) Flavonoid profiling in leaves of Citrus genotypes under different environmental situations. J Agric Food Chem 56:11087–11097
7. Kesterson JW, Hendrickson R (1957) Naringin, a bitter principle of grapefruit. Univ Fla Exp Stn Bull #511A
8. Hendrickson R, Kesterson JW (1964) Hesperidin in florida oranges. Univ Fla Exp Stn Bull #684

9. McIntosh CA, Mansell RL (1997) Three-dimensional analysis of limonin, limonoate a-ring monolactone, and naringin in the fruit of three varieties of *Citrus paradisi*. J Agric Food Chem 45:2876–2883

10. Mansell RL, McIntosh CA, Vest SE (1983) An analysis of the limonin and naringin content of grapefruit juice samples collected from florida state test houses. J Agric Food Chem 31:156–162

11. Barthe GA, Jourdan PS, McIntosh CA, Mansell RL (1988) Radioimmunoassay for the quantitative determination of hesperidin in *Citrus sinensis*. Phytochemistry 27:249–254

12. Jourdan PS, McIntosh CA, Mansell RL (1985) Naringin levels in Citrus tissues II. Quantitative distribution of naringin in *Citrus paradisi* Macfad. Plant Physiol 77:903–908

13. Moriguchi T, Kita M, Tomono Y, Endo-Inagaki T, Omura M (2008) Gene expression in flavonoid biosynthesis: correlation with flavonoid accumulation in developing citrus fruit. Physiol Plant 111:66–74

14. Barthe GA, Jourdan PS, McIntosh CA, Mansell RL (1987) Naringin and limonin production in callus cultures and regenerated plants from *Citrus* sp. J Plant Physiol 127:55–65

15. Mansell RL, McIntosh CA (1991) *Citrus* spp.: in vitro culture and the production of naringin and limonin. In: Bajaj YPS (ed) Biotechnology of medicinal and aromatic plants, vol. 3. Springer, New York, NY, pp 193–210

16. Koca U, Berhow M, Febres V, Champ K, Carillo-Mendoza O, Moore G (2009) Decreasing unpalatable flavonoid components in citrus: the effect of transformation construct. Physiol Plant. doi:10.1111/j.1399- 3054.2009.01264.x

17. Neish AC (1960) Biosynthetic pathways of aromatic compounds. In: Machlis L, Briggs WR (eds) Annual review of plant physiology, vol. 11. Annual Reviews Inc, Palo Alto, CA, pp 55–80

18. Koukol J, Conn EE (1961) The metabolism of aromatic compounds in higher plants. IV. Purification and properties of the phenylalanine deaminase of Hordeum vulgare. J Biol Chem 236:2692–2698

19. Watts KT, Mijts BN, Lee PC, Manning AJ, Schmidt-Dannert C (2006) Discovery of a substrate selectivity switch in tyrosine ammonia-lyase, a member of the aromatic amino acid lyase family. Chem Biol 13(12):1317–1326

20. Hanson KR, Havir EA (1979) An introduction to the enzymology of phenylpropanoid biosynthesis. In: Swain T et al (eds) Recent advances in phytochemistry, vol. 12. Plenum Press, New York, NY, pp 91–137

21. Heller W, Forkmann G (1994) Biosynthesis of flavonoids. In: Harborne JB (ed) The flavonoids: advances in research since 1986. Chapman and Hall, London, pp 499–536

22. Calabrese JC, Jordan DB, Boodhoo A, Sariaslani S, Vannelli T (2004) Crystal structure of phenylalanine ammonia lyase: multiple helix dipoles implicated in catalysis. Biochemistry 43(36):11403–11416

23. Ritter H, Schulz GE (2004) Structural basis for the entrance into the phenylpropanoid metabolism catalyzed by phenylalanine ammonia-lyase. Plant Cell 16(12):3426–3436

24. Riov J, Monselise SP, Kahan RS (1968) Effect of γ-radiation on PAL activity and accumulation of phenolic compounds in Citrus fruit. Radiat Bot 8:463–466

25. Maier VP, Hasegawa S (1970) L-Phenylalanine ammonia-lyase activity and naringenin glycoside accumulation in developing grapefruit. Phytochemistry 9:139–144

26. Thorpe TA, Maier VP, Hasegawa S (1971) Phenylalanine ammonia-lyase activity in citrus fruit tissue cultured in vitro. Phytochemistry 10:711–718

27. Riov J, Monselise SP, Kahan RS (1969) Ethylene-controlled induction of phenylalanine ammonia-lyase in citrus fruit peel. Plant Physiol 44(5):631–635

28. Sanchez-Ballesta MT, Lafuente MT, Zacarias L, Granell A (2000) Involvement of phenylalanine ammonia-lyase in the response of fortune mandarin fruits to cold temperature. Physiol Plant 108(4):382–389

29. Lucheta AR, Silva-Pinhati ACO, Basilio-Palmieri AC, Berger IJ, Freitas-Astua J, Cristofani M (2007) An *in silico* analysis of the key genes involved in flavonoid biosynthesis in *Citrus sinensis*. Genet Mol Biol 30(3):819–831

30. Cramer CL, Edwards K, Dron M, Liang X, Dildine SL, Bolwell GP, Dixon R, Lamb CJ, Schuch W (1989) Phenylalanine ammonia lyase gene organization and structure. Plant Mol Biol 12:367–383

31. Cochrane FC, Davin LB, Lewis NG (2004) The arabidopsis phenylalanine ammonia lyase gene family: kinetic characterization of the four PAL isoforms. Phytochemistry 65(11):1557–1564

32. Sanchez-Ballesta MT, Zacarias L, Granell A, Lafuente MT (2000) Accumulation of PAL transcript and PAL activity as affected by heat-conditioning and low-temperature storage and its relation to chilling sensitivity in mandarin fruits. J Agric Food Chem 48(7):2726–2731

33. Lafuente MT, Zacarias L, Martinez-Tellez MA, Sanchez-Ballesta MT, Dupille E (2001) Phenylalanine ammonia-lyase as related to ethylene in the development of chilling symptoms during cold storage of citrus fruits. J Agric Food Chem 49(12):6020–6025

34. Lafuente MT, Sala JM, Zacarias L (2004) Active oxygen detoxifying enzymes and phenylalanine ammonia-lyase in the ethylene-induced chilling tolerance in citrus fruit. J Agric Food Chem 52(11):3606–3611

35. Oufedjikh H, Mahrouz M, Amiot MJ, Lacroix M (2000) Effect of gamma-irradiation on phenolic compounds and phenylalanine ammonia-lyase activity during storage in relation to peel injury from peel of *Citrus clementina* Hort. ex. Tanaka. J Agric Food Chem 48(2): 559–565

36. Chapple C (1998) Molecular-genetic analysis of plant cytochrome P450-dependent monooxygenases. In: Jones RL et al (eds) Annual review of plant physiology and plant molecular biology, vol. 49. Annual Reviews, Palo Alto, CA, pp 311–343

37. Russell DW, Conn EE (1967) The cinnamic 4-hydroxylase of pea seedlings. Arch Biochem Biophys 122:256–258

38. Russell D (1971) The metabolism of aromatic compounds in higher plants: X. Properties of the cinnamic 4-hydroxylase of pea seedlings and some aspects of its metabolic and experimental control. J Biol Chem 246(12):3870–3878

39. Gabriac B, Werck-Reichhart D, Teutsch H, Durst F (1991) Purification and immunocharacterization of a plant cytochrome P450: the cinnamic acid 4-hydroxylase. Arch Biochem Biophys 288(1):302–309

40. Mizutani M, Ohta D, Sato R (1993) Purification and characterization of a cytochrome P450 (*trans*-cinnamic acid 4-hydroxylase) from etiolated mung bean seedlings. Plant Cell Physiol 34:481–488

41. Teutsch HG, Hasenfratz MP, Lesot A, Stoltz C, Garnier JM, Jeltsch JM, Durst F, Werck-Reichhart D (1993) Isolation and sequence of a cDNA encoding the Jerusalem artichoke cinnamate 4-hydroxylase, a major plant cytochrome P450 involved in the general phenylpropanoid pathway. Proc Natl Acad Sci USA 90(9):4102–4106

42. Fahrendorf T, Dixon RA (1993) Stress responses in alfalfa (*Medicago sativa* L.) XVIII: Molecular cloning and expression of the elicitor-inducible cinnamic acid 4-hydroxylase cytochrome P450. Arch Biochem Biophys 305(2):509–515

43. Ro DK, Mah N, Ellis BE, Douglas CJ (2001) Functional characterization and subcellular localization of poplar (*Populus trichocarpa* x *Populus deltoides*) cinnamate 4-hydroxylase. Plant Physiol 126(1):317–329

44. Winkel-Shirley B (1999) Evidence for enzyme complexes in the phenylpropanoid and flavonoid pathways. Physiol Plant 107:142–149

45. Betz C, McCollum TG, Mayer RT (2001) Differential expression of two cinnamate 4-hydroxylase genes in 'valencia' orange (*Citrus sinensis* Osbeck) Plant Mol Biol 46(6): 741–748

46. Hahlbrock K, Grisebach H (1970) Formation of coenzyme a esters of cinnamic acids with an enzyme preparation from cell suspension cultures of parsley. FEBS Lett 11(1):62–64

47. Hamberger B, Hahlbrock K (2004) The 4-coumarate:CoA ligase gene family in *Arabidopsis thaliana* comprises one rare, sinapate-activating and three commonly occurring isoenzymes. Proc Natl Acad Sci USA 101(7):2209–2214

48. Heller W, Forkmann G (1988) Biosynthesis. In: Harborne JB (ed) The flavonoids: advances in research since 1980. Chapman and Hall, London, pp 399–425

49. Kreuzaler F, Hahlbrock K (1972) Enzymatic synthesis of aromatic compounds in higher plants: formation of naringenin (5,7,4'-trihydroxyflavanone) from p-coumaroyl coenzyme A and malonyl coenzyme A. FEBS Lett 28(1):69–72

50. Kreuzaler F, Mahlberook K (1975) Enzymatic synthesis of an aromatic ring from acetate unit: parital purification and some properties of flavanone synthase from cell suspension cultures of *Petroselinium hortense*. Eur J Biochem 56:205–213

51. Ferrer JL, Jez JM, Bowman ME, Dixon RA, Noel JP (1999) Structure of chalcone synthase and the molecular basis of plant polyketide biosynthesis. Nat Struct Biol 6(8):775–784

52. Austin MB, Noel JP (2003) The chalcone synthase superfamily of type III polyketide synthases. Nat Prod Rep 20(1):79–110

53. Reimold U, Kroger M, Kreuzaler F, Hahlbrock K (1983) Coding and 3' non-coding nucleotide sequence of chalcone synthase mRNA and assignment of amino acid sequence of the enzyme. EMBO J 2(10):1801–1805

54. Schijlen EWGM, de Vos CHR, Martens S, Jonker HH, Rosin FM, Molthoff JW, Tikunov YM, Angenent GC, van Tunen AJ, Bovy AG (2007) RNA interference silencing of chalcone synthase, the first step in the flavonoid biosynthesis pathway, leads to parthenocarpic tomato fruits. Plant Physiol 144:1520–1530

55. Lewinsohn E, Britsch L, Mazur Y, Gressel J (1989) Flavanone glycoside biosynthesis in *Citrus*: chalcone synthase, UDP-glucose:flavanone-7-O-glucosy-transferase and -rhamnosyltransferase activities in cell-free extracts. Plant Physiol 91:1321–1328

56. Moriguchi T, Kita M, Tomono Y, EndoInagaki T, Omura M (1999) One type of chalcone synthase gene expressed during embryogenesis regulates the flavonoid accumulation in citrus cell cultures. Plant Cell Physiol 40(6):651–655

57. Moriguchi T, Kita M, Tomono Y, Endo-Inagaki T, Omura M (2001) Gene expression in flavonoid biosynthesis: correlation with flavonoid accumulation in developing citrus fruit. Physiol Plant 111(1):66–74

58. Lu X, Zhou W, Gao F (2009) Cloning, characterization and localization of CHS gene from blood orange, *Citrus sinensis* (L.) Osbeck cv. Ruby. Mol Biol Rep 36(7):1983–1990

59. Koca U (2007) Elevation of the flavonoid content in grapefruit by introducing chalcone isomerase gene via biotechnological methods. Turk J Pharm Sci 4(3):115–124

60. Bednar RA, Hadcock JR (1988) Purification and characterization of chalcone isomerase from soybeans. J Biol Chem 263(20):9582–9588

61. Moustafa E, Wong E (1967) Purification and properties of chalcone-flavanone isomerase from soya bean seed. Phytochemistry 6:625–632

62. Jez JM, Bowman ME, Dixon RA, Noel JP (2000) Structure and mechanism of the evolutionarily unique plant enzyme chalcone isomerase. Nat Struct Biol 7(9):786–791

63. Shimokoriyama M (1957) Interconversion of chalcones and flavanones of phloroglucinol-type structure. J Am Chem Soc 79(15):4199–4202

64. Raymond WR, Maier VP (1977) Chalcone cyclase and flavonoid biosynthesis in grapefruit. Phytochemistry 16:1535–1539

65. McIntosh C, Mansell R (1990) Biosynthesis of naringin in *Citrus paradisi*: UDP-glucosyltransferase activity in grapefruit seedlings. Phytochemistry 29:1533–1538

66. Fouche SD, Dubery IA (1994) Chalcone isomerase from *Citrus sinensis*: purification and characterization. Phytochemistry 37:127–132

67. Britsch L, Heller W, Grisebach. H (1981) Conversion of flavanone to flavone, dihydroflavonol and flavonol with an enzyme system from cell cultures of parsley. Z Naturforsch 36C:742–750

68. Britsch L, Grisebach H (1986) Purification and characterization of (2S)-flavanone 3-hydroxylase from *Petunia hybrida*. Eur J Biochem 156(3):569–577

69. Britsch L, Dedio J, Saedler H, Forkmann G (1993) Molecular characterization of flavanone 3 beta-hydroxylases. Consensus sequence, comparison with related enzymes and the role of conserved histidine residues. Eur J Biochem 217(2):745–754
70. Lukacin R, Britsch RLL (1997) Identification of strictly conserved histidine and arginine residues as part of the active site in *Petunia hybrida* flavanone 3β-hydroxylase. Eur J Biochem 249:748–757
71. Lukacin R, Groning I, Schiltz E, Britsch L, Matern U (2000) Purification of recombinant flavanone 3β-hydroxylase from *Petunia hybrida* and assignment of the primary site of proteolytic degradation. Arch Biochem Biophys 375(2):364–370
72. Wellmann F, Matern U, Lukacin R (2004) Significance of C-terminal sequence elements for petunia flavanone 3β-hydroxylase activity. FEBS Lett 561(1-3):149–154
73. Owens DK, Crosby KC, Runac J, Howard B, Winkel BS (2008) Biochemical and genetic characterization of arabidopsis flavanone 3-hydroxylase. Plant Physiol Biochem 46: 833–843
74. Punyasiri PA, Abeysinghe IS, Kumar V, Treutter D, Duy D, Gosch C, Martens S, Forkmann G, Fischer TC (2004) Flavonoid biosynthesis in the tea plant *Camellia sinensis*: properties of enzymes of the prominent epicatechin and catechin pathways. Arch Biochem Biophys 431(1):22–30
75. Halbwirth H, Fischer TC, Schlangen K, Rademacher W, Schleifer K, Forkmann G, Stitch K (2006) Screening for inhibitors of 2-oxoglutarate-dependent dioxygenases: flavanone 3-hydroxylase and flavonol synthase. Plant Sci 171:194–205
76. Owens DK, Hale T, Wilson LJ, McIntosh CA (2002) Quantification of the production of dihydrokaempferol by flavanone 3-hydroxytransferase using capillary electrophoresis. Phytochem Anal 13(2):69–74
77. Pelt JL, Downes WA, Schoborg RV, McIntosh CA (2003) Flavanone 3-hydroxylase expression in *Citrus paradisi* and *Petunia hybrida* seedlings. Phytochemistry 64(2):435–444
78. Sutter A, Poulton J, Grisebach H (1975) Oxidation of flavanone to flavone with cell-free extracts from young parsley leaves. Arch Biochem Biophys 170(2):547–556
79. Martens S, Forkmann G, Matern U, Lukacin R (2001) Cloning of parsley flavone synthase I. Phytochemistry 58:43–46
80. Stotz G, Forkmann G (1981) Oxidation of flavanones to flavones with flower extracts of *Antirrhinum majus* (snapdragon). Z Naturforsch 36C:737–741
81. Martens S, Mithofer A (2005) Flavones and flavone synthases. Phytochemistry 66(20):2399–2407
82. Yun CS, Yamamoto T, Nozawa A, Tozawa Y (2008) Expression of parsley flavone synthase I establishes the flavone biosynthetic pathway in *Arabidopsis thaliana*. Biosci Biotechnol Biochem 72(4):968–973
83. Spribille R, Forkmann G (1984) Conversion of dihydroflavonols to flavonols with enzyme extracts from flower buds of *Matthiola incana*. Z Naturforsch 39C:714–719
84. Stitch K, Eidenberger T, Wurst F, Forkmann G (1992) Flavonol synthase activity and the regulation of flavonol and anthocyanin biosynthesis during flower development in *Dianthus caryophyllus*. Z Naturforsch 47C:553–560
85. Holton TA, Brugliera F, Tanaka Y (1993) Cloning and expression of flavonol synthase from *Petunia hybrida*. Plant J 4(6):1003–1010
86. Pelletier MK, Murrell JR, Shirley BW (1997) Characterization of flavonol synthase and leucoanthocyanidin dioxygenase genes in arabidopsis. Further evidence for differential regulation of "early" and "late" genes. Plant Physiol 113(4):1437–1445
87. Owens DK, Alerding AB, Crosby KC, Bandara AB, Westwood JH, Winkel BS (2008) Functional analysis of a predicted flavonol synthase gene family in arabidopsis. Plant Physiol 147(3):1046–1061
88. Lukacin R, Britsch L (1997) Identification of strictly conserved histidine and arginine residues as part of the active site in *Petunia hybrida* flavanone 3β-hydroxylase. Eur J Biochem 249(3):748–757

89. Wilmouth RC, Turnbull JJ, Welford RW, Clifton IJ, Prescott AG, Schofield CJ (2002) Structure and mechanism of anthocyanidin synthase from *Arabidopsis thaliana*. Structure 10(1):93–103

90. Stracke R, De Vos RC, Bartelniewoehner L, Ishihara H, Sagasser M, Martens S, Weisshaar B (2009) Metabolomic and genetic analyses of flavonol synthesis in *Arabidopsis thaliana* support the in vivo involvement of leucoanthocyanidin dioxygenase. Planta 229(2):427–445

91. Preuss A, Stracke R, Weisshaar B, Hillebrecht A, Matern U, Martens S (2009) *Arabidopsis thaliana* expresses a second functional flavonol synthase. FEBS Lett 583(12):1981–1986

92. Moriguchi T, Kita M, Ogawa K, Tomono Y, Endo T, Omura M (2002) Flavonol synthase gene expression during citrus fruit development. Physiol Plant 114(2):251–258

93. Wellmann F, Lukacin R, Moriguchi T, Britsch L, Schiltz E, Matern U (2002) Functional expression and mutational analysis of flavonol synthase from *Citrus unshiu*. Eur J Biochem 269(16):4134–4142

94. Prescott AG, Stamford NPJ, Wheeler G, Firmin JL (2002) In vitro properties of a recombinant flavonol synthase from *Arabidopsis thaliana*. Phytochemistry 60:589–593

95. Lukacin R, Wellmann F, Britsch L, Martens S, Matern U (2003) Flavonol synthase from *Citrus unshiu* is a bifunctional dioxygenase. Phytochemistry 62(3):287–292

96. Stafford HA, Lester HH (1982) Enzymic and nonenzymic reduction of (+)-dihydroquercetin to its 3,4,-diol. Plant Physiol 70(3):695–698

97. Heller W, Britsch L, Forkmann G, Grisebach H (1985) Leucoanthocyanidins as intermediates in anthocyanin biosynthesis in flowers of *Matthiola incana*. R. Br. Planta 163(2):191–196

98. Heller W, Forkmann G, Britsch L, Grisebach H (1985) Enzymatic reduction of (+)-dihydroflavonols to flavan-3,4-*cis*-diols with flower extracts from *Matthiola incana* and its role in anthocyanin biosynthesis. Planta 165(2):284–287

99. Fischer D, Stich K, Britsch L, Grisebach H (1988) Purification and characterization of (+)dihydroflavonol (3-hydroxyflavanone) 4-reductase from flowers of *Dahlia variabilis*. Arch Biochem Biophys 264(1):40–47

100. Yu O, Matsuno M, Subramanian S (2006) Flavonoid compounds in flowers: genetics and biochemistry. In: da Silva JAT (ed) Floriculture, ornamental and plant biotechnology, vol. 1. Global Sciences Book, London, pp 282–292

101. Lo Piero AR, Puglisi I, Rapisarda P, Petrone G (2005) Anthocyanins accumulation and related gene expression in red orange fruit induced by low temperature storage. J Agric Food Chem 53(23):9083–9088

102. Lo Piero AR, Puglisi I, Petrone G (2006) Gene characterization, analysis of expression and in vitro synthesis of dihydroflavonol 4-reductase from [*Citrus sinensis* (L.) Osbeck]. Phytochemistry 67(7):684–695

103. Reddy GM, Coe EH Jr (1962) Inter-tissue complementation: a simple technique for direct analysis of gene-action sequence. Science 138(3537):149–150

104. Menssen A, Hohmann S, Martin W, Schnable PS, Peterson PA, Saedler H, Gierl A (1990) The En/Spm transposable element of *Zea mays* contains splice sites at the termini generating a novel intron from a dSpm element in the A2 gene. EMBO J 9(10):3051–3057

105. Saito K, Kobayashi M, Gong Z, Tanaka Y, Yamazaki M (1999) Direct evidence for anthocyanidin synthase as a 2-oxoglutarate-dependent oxygenase: molecular cloning and functional expression of cDNA from a red form of *Perilla frutescens*. Plant J 17(2):181–189

106. Turnbull JJ, Sobey WJ, Aplin RT, Hassan A, Firmin JL, Schofield CJ, Prescott AG (2000) Are anthocyanidins the immediate products of anthocyanidin synthase? Chem Commun 24:2473–2474

107. Martens S, Forkmann G, Britsch L, Wellmann F, Matern U, Lukacin R (2003) Divergent evolution of flavonoid 2-oxoglutarate-dependent dioxygenases in parsley. FEBS Lett 544 (1–3):93–98

108. Turnbull JJ, Nakajima J, Welford RW, Yamazaki M, Saito K, Schofield CJ (2004) Mechanistic studies on three 2-oxoglutarate-dependent oxygenases of flavonoid biosynthesis: anthocyanidin synthase, flavonol synthase, and flavanone 3β-hydroxylase. J Biol Chem 279(2):1206–1216

109. Cotroneo PS, Russo MP, Ciuni M, Recupero GR (2006) Quantitative real-time reverse transcriptase-PCR profiling of anthocyanin biosynthetic genes during orange fruit ripening. J Amer Soc Hort Sci 131(4):537–543

110. Fritsch H, Grisebach. H (1975) Biosynthesis of cyanidin in cell cultures of *Haplopappus gracillus*. Phytochemistry 14:2437–2442

111. Hagmann ML, Heller W, Grisebach H (1983) Induction and characterization of a microsomal flavonoid 3′-hydroxylase from parsley cell cultures. Eur J Biochem 134(3):547–554

112. Brugliera F, Barri-Rewell G, Holton TA, Mason JG (1999) Isolation and characterization of a flavonoid 3′-hydroxylase cDNA clone corresponding to the Ht1 locus of *Petunia hybrida*. Plant J 19(4):441–451

113. Doostdar H, Shapiro JP, Niedz R, Burke MD, McCollum TG, McDonald RE, Mayer RT (1995) A cytochrome P450 mediated naringenin 3′-hydroxylase from sweet orange cell cultures. Plant Cell Physiol 36(1):69–77

114. Stotz G, Forkmann G (1982) Hydroxylation of the B-ring of flavonoids in the 3′- and 5′-position with enzyme extracts from the flowers of *Verbena hybrida*. Z Natuforsch 37C:19–23

115. Hertog MGL, Hollman PCH, van de Putte B (1993) Content of potentially anticarcinogenic flavonoids of tea infusions, wines, and fruit juices. J Agric Food Chem 41:1242–1246

116. Maccarone E, Maccarone A, Perrini G, Rapisarda P (1983) Anthocyanins of the moro orange juice. Ann Chim 73:533–539

117. Maccarone E, Maccarone A, Rapisarda P (1985) Acylated anthocyanins from orange. Ann Chim 75:79–86

118. Dugo P, Mondello L, Morabito D, Dugo G (2003) Characterization of the anthocyanin fraction of sicilian blood orange juice by micro-HPLC-ESI/MS. J Agric Food Chem 51(5):1173–1176

119. Hillebrand S, Schwarz M, Winterhalter P (2004) Characterization of anthocyanins and pyranoanthocyanins from blood orange [*Citrus sinensis* (L.) Osbeck] juice. J Agric Food Chem 52(24):7331–7338

120. Jay M, Viricel MR, Gonnett JF (2006) C-Glycosylflavonoids. In: Anderson DM, Markham KR (eds) Flavonoids: chemistry, biochemistry, and applications. CRC Press, Boca Raton, FL, pp 857–915

121. Lewinsohn E, Berman B, Mazur Y, Gressel J (1986) Glucosylation of exogenous flavanones by grapefruit (*Citrus paradisi*) cell cultures. Phytochemistry 25:2531–2535

122. Lewinsohn E, Berman E, Mazur Y, Gressel J (1989) (7) Glucosylation and (1-6) rhamnosylation of exogenous flavanones by undifferentiated *Citrus* cell cultures. Plant Sci 61:23–28

123. McIntosh C, Latchinian. L, Mansell R (1990) Flavanone-specific 7-O-glucosyltransferase activity in *Citrus paradisi* seedlings: purification and characterization. Arch Biochem Biophys 282:50–57

124. Berhow M, Smolensky D (1995) Developmental and substrate specificity of hesperetin-7-O-glucosyltransferase activity in *Citrus limon* tissues using high performance liquid chromatographic analysis. Plant Sci 112:139–147

125. Durren RL, McIntosh CA (1999) Flavanone 7-O-glucosyltransferase activity from *Petunia hybrida*. Phytochemistry 52:793–798

126. Owens DK, McIntosh CA (2009) Identification, recombinant expression, and biochemical characterization of a flavonol 3-O-glucosyltransferase clone from *Citrus paradisi*. Phytochemistry 70:1382–1391

127. Jourdan PS, Mansell RL (1982) Isolation and partial characterization of three glucosyl transferases involved in the biosynthesis of flavonol triglucosides in *Pisum sativum* L. Arch Biochem Biophys 213:434–443

128. Kleinehollenhorst G, Behrens H, Pegels G, Srunk N, Weirmann R (1982) Formation of flavonol 3-O-diglycosides and flavonol 3-O-triglycosides by enzyme extracts from anthers of *Tulipa* cv. Apeldorn: characterization and activity of three different O-glycosyltransferases during anther development. Z Naturforsch 37:587–599

129. Bar-Peled M, Lewinsohn E, Fluhr R, Gressel J (1991) UDP-rhamnose:flavanone-7-O-glucoside-2″-O-rhamnosyltransferase: purification and characterization of an enzyme catalyzing the production of bitter compounds in Citrus. J Biol Chem 266:20953–20959

130. Frydman A, Weisshaus O, Bar-Peled M, Huhman D, Summer L, Marin F, Lewinsohn E, Fluhr R, Gressel J, Eyal Y (2004) Citrus fruit bitter flavors: isolation and functional characterization of the gene *Cm1,2RhaT* encoding a 1,2rhamnosyltransferase, a key enzyme in the biosynthesis of the bitter flavonoids of Citrus. Plant J 40:88–100

131. Roysarkar T, Strong CL, Sibhatu MB, Pike LM, McIntosh CA (2007) Cloning, expression, and characterization of a putative flavonoid glucosyltransferase from grapefruit (*Citrus paradisi*) leaves. In: Nikolau BJ, Wurtele ES (eds) Concepts in plant metabolomics. Springer Publishing, Dordrecht, pp 247–259

132. Knisley D, Seier E, Lamb D, Owens D, McIntosh C (2009) A graph-theoretic model based on primary and predicted secondary structure reveals functional specificity in a set of plant secondary product UDP-glucosyltransferases. In: Loging W, Doble M, Sun Z, Malone J (eds) Proceedings of the 2009 international conference on bioinformatics, computational biology, genomics and chemoinformatics (BCBGC-09). ISBN: 978-1-60651-009-4, pp 65–72

133. Tanner DC (2000) Structural elucidation of flavanone 7-O-glucosyltransferase in grapefruit (*Citrus paradisi*) seedlings. M.A. Thesis, East Tennessee State University

134. Osmani SA, Bak S, Moller BL (2009) Substrate specificity of plant UDP-dependent glucosyltransferases predicted from crystal structures and homology modeling. Phytochemistry 70:325–347

135. Kalinina OV, Gelfand MS, Russell RB (2009) Combining specificity determining and conserved residues improves functional site prediction. Bioinformatics. doi:10.1186/1471–2105–10–174

136. Kita M, Hirata Y, Moriguchi T, Endo-Inagaki T, Matsumoto R, Hasegawa S, Sunhayda CG, Omura M (2000) Molecular cloning and characterization of a novel gene encoding limonoid UDP-glucosyltransferase in Citrus. FEBS Lett 469:173–178

137. Ford CM, Boss PK, Hoj PB (1998) Cloning and characterization of *Vitis vinifera* UDP-glucose:flavonoid 3-O-glucosyltransferase, a homologue of the enzyme encoded by the maize bronze-1 locus that may primarily serve to glucosylate anthocyanidins in vivo. J Biol Chem 273:9224–9233

138. Shao H, He X, Achnine L, Blount JW, Dixon RA, Wang X (2005) Crystal structures of a multifunctional triterpene/flavonoid glycosyltransferase from *Medicago truncatula*. Plant Cell 17:3141–3154

139. Offen W, Martinez-Fleites C,, Yang M, Kiat-Lim E, Davis BG, Tarling CA, Ford C, Bowles DJ, Davies GJ (2006) Structure of a flavonoid glucosyltransferase reveals the basis for plant natural product modification. EMBO J 25:1396–1405

140. Ogata J, Itoh Y, Ishida M, Yoshida Y, Ozeki Y (2004) Cloning and heterologous expression of cDNAs encoding flavonoid glucosyltransferases from *Dianthus caryophyllus*. Plant Biotechnol 21:367–375

141. Kim JH, Park Y, Ko JH, Lim CE, Lim J, Lim Y, Ahn JH (2006) Characterization of flavonoid 7-O-glucosyltransferase from *Arabidopsis thaliana*. Biosci Biotechnol Biochem 70: 1471–1477

142. Do CB, Cormier F, Nicolas Y (1995) Isolation and characterization of a UDP-glucose:cyanidin 3-O-glucosyltransferase from grape cell suspension cultures (*Vitis vinifera* L.). Plant Sci 112:43–51

143. Griesser M, Vitzthum F, Fink B, Bellido ML, Raasch C, Munoz-Blanco J, Schwab W (2008) Multi-substrate flavonol O-glucosyltransferases from strawberry (*Fragaria* x *Ananassa*) achene and receptacle. J Exp Bot. doi:10.1093/jxb/ern117

144. Hartmann T (2007) From waste products to ecochemicals: 50 years research of plant secondary metabolism. Phytochemistry 68:2831–2846
145. Gould KS, Lister C (2006) Flavonoid functions in plants. In: Anderson DM, Markham KR (eds) Flavonoids: chemistry, biochemistry, and applications. CRC Press, Boca Raton, FL, pp 397–442
146. DelRio JA, Gomez P, Baidez AG, Arcas MC, Botia JM, Ortuno A (2004) Changes in the levels of polymethoxyflavones and flavanones as part of the defense mechanism of *Citrus sinensis* (cv. Valencia Late) fruits against *Phytophthora citrophthora*. J Agric Food Chem 52:1813–1917
147. Manthey JA, Grohman K, Berhow MA, Tisserat B (2000) Changes in citrus leaf flavonoid concentrations resulting from blight-induced zinc-deficiency. Plant Physiol Biochem 38:333–343
148. Ortuno A, Baidez A, Gomez P, Arcase MC, Porras I, Garcia-Lidon A, DelRio JA (2006) *Citrus paradisi* and *Citrus sinensis* flavonoids: Their influence in the defence mechanism against *Penicillium digitatum*. Food Chem 98:351–358
149. Clifford MN, Brown JE (2006) Dietary flavonoids and health – broadening the perspective. In: Anderson DM, Markham KR (eds) Flavonoids: chemistry, biochemistry, and applications. CRC Press, Boca Raton, FL, pp 319–370
150. Anderson, DM, Markham KR (eds) (2006) Flavonoids: chemistry, biochemistry, and applications. CRC Press, Boca Raton, FL
151. Passamonti S, Terdoslavich M, Franca R, Vanzo A, Tramer F, Braidot E, Petrussa E, Vianello A (2009) Bioavailability of flavonoids: a review of their membrane transport and the function of bilitranslocase in animal and plant organisms. Curr Drug Metab 10:369–394
152. Bhimanagouda SP, Turner ND, Miller EG, Brodbelt JS (eds) (2006) Potential health benefits of Citrus, ACS symposium series, vol 936
153. Datla KP, Zbarsky V, Rai D, Parkar S, Osakabe N, Aruoma OI, Dexter DT (2007) Short-term supplementation with plant extracts rich in flavonoids protect nigrostriatal dopaminergic neurons in a rat model of Parkinson's disease. J Am Coll Nutr 26:341–349
154. Benavente-Garcia O, Castillo J (2008) Update on uses and properties of Citrus flavonoids: new findings in anticancer, cardiovascular, and anti- inflammatory activity. J Agric Food Chem 56:6185–6205
155. Londono-Londono J, Rodrigues de Lima V, Lara O, Gil A, Pasa TBC, Arango GJ, Pineda JRR (2009) Clean recovery of antioxidant flavonoids from citrus peel: optimizing an aqueous ultrasound-assisted extraction method. J Food Chem. doi:10.1016/j.foodchem.2009.05.075
156. Kuoryanagi M, Ishii JH, Kawahara N, Sugimoto H, Yamada H, Okihara K, Shirota O (2008) Flavonoid glycosides and limonoids from *Citrus* molasses. J Nat Med 62:107–111
157. Talon M, Gmitter FG (2008) Citrus genomics. Int J Plant Genomics Article ID 528361, doi:10.115/2008/528361
158. Seelenfreund D, Chiong M, Lobos S, Perez LM (1996) A full-length cDNA coding for phenylalanine ammonia-lyase from *Citrus limon* (Accession No.U43338) (PGR96-026). Plant Physiol 111:348
159. Deng ZN, Gentile A, La Malfa S, Domina F, Tribulato E, Messina A (2001) Identification of stress resistance genes by PCR-Select cDNA subtraction in Citrus. In Proceedings of the 5th congress of the European foundation for plant pathology I (Societa Italiana di Patologia Vegetale, ed.), pp 261–263
160. Gomi K, Yamamoto H, Akimitsu K (2002) Characterization of a lipoxygenase gene in rough lemon induced by *Alternaria alternate*. J Gen Plant Pathol 68:21–30

Chapter 7
Ginsenoside Variation Within and Between Ontario Ginseng Landraces: Relating Phytochemistry to Biological Activity

Kristina L. McIntyre, Alice Luu, Cathy Sun, Dan Brown, E.M.K. Lui, and John T. Arnason

7.1 Introduction

Cultivated American ginseng (*P. quinquefolius* L., Araliaceae) is one of North America's most valuable medicinal crops, worth an estimated $100 million annually, and is also one of the top-selling dietary supplements in the United States [1, 2]. Ontario has recently become the largest grower of American ginseng with over 5,000 acres in production [3]. Ginseng is commercially important in Canada and the United States and great quantities are exported to Asian markets [3].

In traditional Chinese medicine, American ginseng is used to treat a "yin" deficiency, or to replenish vital energy as a general tonic. Ginseng has been examined for a wide array of activities, including antidiabetic, anticancer, and antioxidant activities, and immunostimulation [4–8] and it contains several bioactive chemical constituents, including ginsenosides, polysaccharides, and polyacetylenes. Ginsenosides have received the most attention as they are commonly believed to be responsible for the majority of ginseng's biological activity. Roots have been the focus of study due to their high ginsenoside levels, traditional use, and high commercial value.

The amount of ginsenosides present in roots varies due to several factors including plant age, time of harvest, and location [9–12]. A recent study of ginseng populations in New York State reported some populations with higher Rg1 than Re content, the opposite of previous reports [10]. An inverse relationship between Rg1 and Re was observed where when Rg1 content was low, Re content was high and when Re content was high, Rg1 content was low. Schlag and McIntosh [9] later identified two distinct chemotypes, a low Rg1/high Re Northern chemotype and a high Rg1/low Re Southern chemotype. Variation between locations is of particular interest due to the potential for unique ginsenoside profile and chemotype

J.T. Arnason (✉)
Department of Biology, University of Ottawa, Ottawa, ON, Canada
e-mail: john.arnason@uottawa.ca

D.R. Gang (ed.), *The Biological Activity of Phytochemicals*, Recent Advances in Phytochemistry 41, DOI 10.1007/978-1-4419-7299-6_7,
© Springer Science+Business Media, LLC 2011

identification. Different ginsenoside profiles may translate to different levels of biological activity. Identifying differences in ginsenoside profiles could be important for quality control and landrace development.

Ontario ginseng cultivation began over 100 years ago with the Hellyer family in Norfolk County [3]. Seeds were cultivated from wild root and cultivation in Ontario has grown to include close to 400 growers. Today Ontario ginseng comprises several unimproved landraces [13, 14], which are farmer-selected populations that have likely diverged significantly from wild populations, developing separately for several decades. This study examines the variation of six major ginsenosides (Fig. 7.1) within and between five Ontario landraces. As ginsenosides have been associated with antidiabetic and antioxidant activities, the effect of ginsenoside variation on these activities is discussed. This is one step toward improving the characterization of Ontario ginseng.

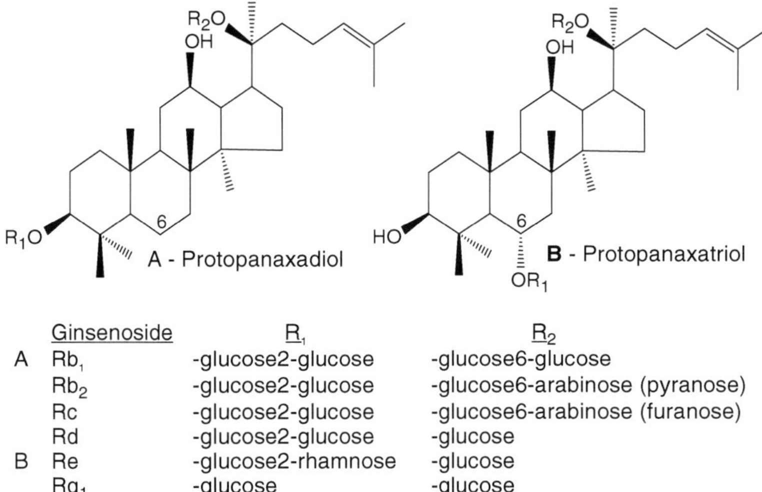

	Ginsenoside	R_1	R_2
A	Rb_1	-glucose2-glucose	-glucose6-glucose
	Rb_2	-glucose2-glucose	-glucose6-arabinose (pyranose)
	Rc	-glucose2-glucose	-glucose6-arabinose (furanose)
	Rd	-glucose2-glucose	-glucose
B	Re	-glucose2-rhamnose	-glucose
	Rg_1	-glucose	-glucose

Fig. 7.1 Chemical structures of six major ginsenosides – (**a**) protopanaxadiol and (**b**) protopanaxatriol

7.2 Methods

7.2.1 Ginseng Collection and Extraction

A total of 21–25 4-year-old roots were collected from 1 m × 1 m plots from five ginseng farms in Norfolk County, Ontario. Root sections with diameter between 4 and 10 mm were examined. A total of 400 mg of ground root was added to 10 mL of 70% methanol and sonicated for 25 min at room temperature, centrifuged and the supernatant removed. The extraction was repeated twice using 10 and 4 mL, phases combined and brought to 25 mL. For HPLC analysis, 100 μL of 5% KOH was added to 1 mL of extract and incubated in the dark for 2 h. The extract was then neutralized with 14% KH_2PO_4 and filtered through a 0.2-μm PTFE filter [15]. For assays, extracts were dried using a speed vac and lyophilized.

7.2.2 HPLC-DAD Analysis

Samples were analyzed using a validated HPLC-DAD method developed by Paula Brown [15] and ginsenosides Rg1, Re, Rb1, Rc, Rb2, and Rd were quantified. HPLC analysis was conducted on a Hewlett-Packard Chemstation series 1100 chromatograph (Agilent, Palo Alto, CA) consisting of an autosampler, quaternary pump, and a diode array detector (DAD). Solvents were HPLC grade (Fisher Scientific, Ottawa, ON). A Phenomenex Luna C18 column (150 mm × 4.6 mm; 5 μm particle size; 100 Å pore size) was kept at 25°C and a flow rate of 1.5 mL/min maintained. The mobile phase consisted of water (A), 80:20 acetonitrile/water (B), and acetonitrile (C). Initial conditions of 76% A: 24% B were held isocratically for 8 min and then changed following linear gradients to 68% A: 32% B in 18 min, 60% A: 40% B in 25 min, and 52% A: 48% B in 42 min. The column was washed with 100% C over the next 11 min and returned to initial conditions. The column was allowed to re-equilibrate at initial conditions for 7 min between samples. 10 μL of sample was injected and the profile was monitored at 203 nm.

7.2.3 Anti-glycation Activity

The elevated formation of advanced glycation end products (AGEs) is an underlying mechanism of diabetic complications. AGE formation is increased under hyperglycemic conditions and they lead to the crosslinking of proteins, and tissue and cellular damage [16]. Reducing AGE formation could be a potential target for reducing diabetic complications. To determine the potential of Ontario ginseng to inhibit AGE formation, we conducted a fluorescence-based *in vitro* assay. This assay was carried out as described by Farsi et al. [17] with modifications [18]. Briefly, bovine serum albumin (BSA) was incubated with a glucose/fructose mixture, both prepared in phosphate buffer, and extract at various concentrations while shaking for 7 days. Positive (quercetin), negative (vehicle only), extract fluorescence, and BSA fluorescence controls as well as a blank treatment with only buffer were included. Fluorescence was measured at excitation and emission wavelengths of 355 and 460 nm, respectively. Percentage of inhibition and IC_{50} values were calculated as previously described [17].

7.2.4 Antioxidant Activity

Oxidative stress is implicated in a range of health concerns, including diabetes mellitus, cancer, and cardiovascular disease among many other degenerative diseases

[19]. As phenolics are well-known antioxidant agents, the Folin-Ciocalteau total phenolics assay was performed to assess phenolic content in ginseng roots. This assay was performed as described by Singleton et al. [20] with modifications [21]. Briefly, extract dissolved in 70% methanol was combined with Folin-Ciocalteau reagent, incubated for 5 min, and then 7.5% $NaHCO_3$ was added. Samples were transferred into microplate wells in triplicate, incubated in darkness for 2 h, and then absorbance read at 725 nm. Samples were blanked against a treatment with only solvent. A standard curve of quercetin was produced and data were expressed as quercetin equivalents.

Free radical scavenging activity was assessed using the DPPH (1, 1-di-phenyl-2-picrylhydrazyl) assay as described by Harbilas et al. [22] with incubation time increased to 65 min. Briefly, 250 μL of 100 μmol/L DPPH dissolved in methanol was added to 40 μL of extract (tested at 5 concentrations) in a microplate well. A standard curve of ascorbic acid (positive control) was included as a reference and all data were blanked against a treatment with only methanol. Absorbance was read with a microplate reader at 517 nm. The inhibitory concentration for 50% scavenging (IC_{50}) of each extract was calculated and compared to the IC_{50} of the ascorbic acid standard curve.

7.2.5 Statistics

Ginsenoside content was expressed as ginsenoside weight relative to dry root weight (% w/w). As ginsenoside content data were not normally distributed, the Kruskal–Wallis test was used to determine differences in ginsenoside content between landraces. Results were considered significant when $p<0.05$. Linear regression was used to assess the correlation between ginsenoside or phenolic content and DPPH activity.

7.3 Results and Discussion

7.3.1 Ginsenoside Variation

A typical chromatogram from one root sample is presented in Fig. 7.2. Significant variation in the content of most ginsenosides was seen between landraces, with Rg1 being the only ginsenoside not differing in quantity between at least two landraces ($p>0.05$) (Table 7.1). These results may reflect different environmental and farmer-initiated selection pressures on plants at individual farms. Total ginsenoside levels were significantly reduced in one landrace. Ginsenoside composition was similar between landraces with the following general trend: Rb1 > Rd $\approx$ Re > Rc > Rg1 > Rb2. This profile is unique in comparison to American ginseng populations in other locations where Rd is typically one of the lowest contributing ginsenosides [9–11] (Table 7.2). The greatest range in total ginsenosides within an individual landrace

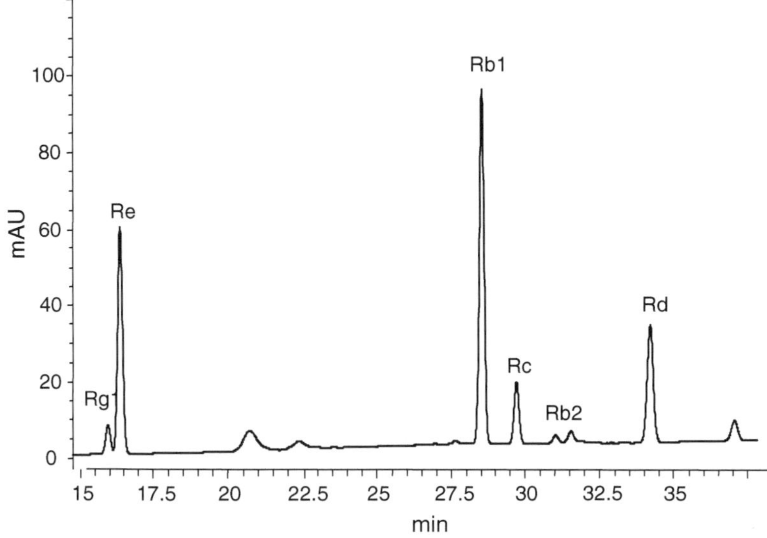

Fig. 7.2 HPLC-DAD chromatogram of one Ontario ginseng root sample

Table 7.1 Mean ginsenoside content (% w/w) (± SD) of roots from 5 Ontario landraces (n = 21–25). Significant differences between populations are indicated by different letters

Landrace	Ginsenoside content (% w/w) (±SE)						
	Rg1	Re	Rb1	Rc	Rb2	Rd	Total
1	0.17 (0.08)[a]	1.45 (0.58)[a]	4.23 (1.36)[a,b]	0.66 (0.28)[a,b]	0.13 (0.05)[a]	1.74 (1.08)[a]	8.38 (2.80)[a]
2	0.17 (0.08)[a]	1.31 (0.41)[a]	4.53 (0.85)[a]	0.58 (0.23)[a,b]	0.11 (0.04)[a,b]	1.83 (1.09)[a]	8.53 (2.11)[a]
3	0.16 (0.09)[a]	1.16 (0.30)[a,b]	4.21 (1.03)[a]	0.44 (0.11)[a]	0.11 (0.07)[b]	1.39 (0.71)[a]	7.47 (1.63)[a]
4	0.18 (0.07)[a]	1.40 (0.47)[a]	4.45 (1.63)[a]	0.63 (0.24)[b]	0.12 (0.04)[a]	1.59 (0.95)[a]	8.38 (2.81)[a]
5	0.17 (0.09)[a]	0.99 (0.27)[b]	3.09 (0.86)[b]	0.46 (0.14)[a]	0.12 (0.08)[a,b]	0.71 (0.32)[b]	5.54 (1.43)[b]

was between 4.36 and 16.61% w/w and the lowest range was between 4.24 and 10.24% w/w (Table 7.3).

Our study showed mean total ginsenoside levels that were greater than means reported in several other locations including New York State, Maryland, British Columbia, and wild populations in Ontario, Quebec, Vermont, Maine, and Wisconsin (Table 7.2) [9–11, 23]. This may be due to the use of root sections in our study versus whole roots. Roots ranging from 5 to 10 mm in diameter have been shown to contain higher levels of some individual ginsenosides than roots with

Table 7.2 Comparison of total ginsenoside content between different locations and cultivation methods

| | Mean ginsenoside content (% w/w) | | | | | | | | |
Study	Total	Rg1	Re	Rb1	Rc	Rb2	Rd	Production method	Location
This study	7.7	0.17	1.26	4.1	0.55	0.12	1.45	Cultivated	Ontario
[23]	3.0	0.18	0.18	1.22	0.18	0.02	0.29	Cultivated	British Columbia
[11]	4.9	0.25	0.25	1.88	0.36	0.13	0.48	Cultivated	Quebec
[11]	5.8	0.94	0.94	2.81	0.42	0.09	0.29	Wild	Ontario, Quebec, Maine, Vermont, Wisconsin
[10]	2.7	0.80	0.80	0.95	0.20	0.05	0.1	Wild and cultivated	New York State
[9] (high Rg1/low Re)	2.3	1.04	1.04	0.69	0.23	n/a	0.05	Wild and cultivated	Maryland
[9] (low Rg1/high Re)	2.5	0.23	0.23	0.83	0.23	n/a	0.1	Wild and cultivated	Maryland

Table 7.3 Range of ginsenoside content (% w/w) within individual landraces ($n = 21-25$)

| | Ginsenoside content range (min.–max. % w/w) | | | | | | |
Landrace	Rg1	Re	Rb1	Rc	Rb2	Rd	Total
1	0.07–0.43	0.66–3.31	2.63–7.47	0.19–1.49	0[a]–0.17	0.42–3.65	5.34–13.74
2	0.09–0.37	0.52–2.57	3.28–6.70	0.30–1.40	0.06–0.24	0.48–4.74	5.39–11.80
3	0.05–0.45	0.67–1.84	2.33–6.89	0.29–0.64	0.06–0.32	0.36–3.33	4.24–10.24
4	0.10–0.35	0.83–2.14	1.97–9.84	0.33–1.32	0.08–0.26	0.70–5.03	4.36–16.61
5	0.10–0.41	0.60–1.73	1.66–5.37	0.23–0.95	0.07–0.40	0.27–1.36	3.03–9.61

[a]Not present at a detectable level.

a 15–38 mm diameter, though total ginsenoside levels were not significantly different between roots within this entire diameter range [24]. It is also possible that differences in extraction methods, chromatographic conditions, and root age could account for the differences in ginsenoside levels.

More important than overall mean ginsenoside levels, it is most significant that high variation is seen between landraces in this study as well as between populations when compared to previous reports. A study of wild American ginseng populations collected in Ontario, Quebec, Maine, Vermont, and Wisconsin and one cultivated population in Quebec showed total ginsenoside levels significantly differed among populations with the highest content at $10.93 \pm 1.96\%$ w/w and lowest at $2.73 \pm 0.63\%$ w/w [11]. This range was greater than the range of total ginsenosides observed between landraces in the current study. This may be due to the wider age range and geographical distribution which would result in greater environmental variability. Furthermore, wild populations are expected to contain more genetic variation than cultivated landraces. As was seen in this study, significant variation was also reported among and within ginseng populations in Maryland with variation in total ginsenosides ranging from 0.85 to 5.78% w/w in wild roots and 1.04–4.07% w/w in cultivated roots [9]. In a study of cultivated ginseng in British Columbia, significant variations were seen among 9 fields, though Rb1, Rc, and Rd were the only individual ginsenosides that varied significantly [23]. The variation in ginsenoside levels between growing locations is important for quality-control purposes as it cannot be assumed that American ginseng obtained from different locations will contain similar ginsenoside content. Furthermore, as levels can vary significantly within populations, it is important to assess multiple roots from each population to obtain an accurate estimation of ginsenoside content.

7.3.2 Anti-glycation and Antioxidant Activity of Ginseng In Vitro

One reason for assessing ginsenoside variation is that variation in active constituents may lead to variation in biological activity. For example, in clinical trials with both non-diabetic participants and patients with well-controlled type 2 diabetes, American ginseng was shown to significantly reduce post-prandial glycemia [7, 25, 26]. However, in a subsequent trial no significant reduction was found and this ginseng batch was shown to have a 48% lower total ginsenoside content compared to the initial effective batch [27]. This demonstrates the importance of considering the phytochemical profile when assessing biological activity.

Ontario ginseng root extract was not a significant inhibitor of AGE formation. Inhibition did not reach 50%, with an extrapolated IC_{50} of >10 mg/mL (Table 7.4). This value was high in comparison to extracts high in phenolic content such as lowbush blueberry and corn silk, which have IC_{50} values in the 1.3–78.7 μg/mL range [17, 18] and the standard quercetin. Pure ginsenosides also had relatively low activity.

Table 7.4 Inhibitory concentration 50% (IC_{50}) of advanced glycation end product formation by pure ginsenosides and an Ontario ginseng extract ($n = 3$)

Ginsenoside	IC_{50} range
Rb1	1–10 mg/mL
Rd	1–10 mg/mL
Re	>10 mg/mL
Rg1	0.5–1.0 mg/mL
Ontario ginseng extract	>10 mg/mL
Quercetin	1.8 μg/mL
Lowbush blueberry leaves	6.3 μg/mL [17]
Lowbush blueberry stem	3.1 μg/mL [17]

Ontario-grown American ginseng extracts were shown to have low total phenolic content in comparison to some previously tested extracts. Ontario ginseng samples ranged from 0.67 to 6.98 μg quercetin equivalents/mg extract whereas values were in the order of 60–269 μg quercetin equivalents/mg extract for plants known to be rich in phenolic compounds (such as quercetin, catechin, and other flavonoids) [21].

Ontario ginseng samples produced IC_{50} values ranging from 1.35 to 3.01 mg/mL in the DPPH radical scavenging assay. The biologically active concentration range was similar to a previously tested American ginseng extract [28]. IC_{50} values of samples were assessed in comparison to the IC_{50} for ascorbic acid and the relative activity was rather low. The maximum value for log IC_{50} ascorbic acid/log IC_{50} ginseng sample was 0.14, which was low in comparison to other samples high in phenolic content that range between 0.5 and 1.1 [22]. There was a correlation between DPPH radical scavenging activity and ginsenoside content (R^2=0.84) (Fig. 7.3a) and also a correlation between DPPH activity and total phenolic content (R^2=0.89) (Fig. 7.3b). In the case of ginseng, the phenolic component is small but likely contributes to the antioxidant activity. Although ginsenoside levels are high, as illustrated above, in general, phenolic-rich plants have tended to have greater antioxidant and anti-glycation activity than ginseng samples. It is possible that the antioxidant activity seen may be more related to phenolics than to ginsenoside content.

Given the literature on anti-diabetic effects of ginseng, it was interesting that Ontario ginseng did not show strong anti-glycation and antioxidant effects in vitro. It may be possible that ginseng acts by influencing glutathione in cells. Oxidative stress can lead to a reduction in glutathione levels and a study of ginsenoside Re administration to diabetic rats resulted in increased glutathione (GSH) levels in the eye and kidney [29]. In a study examining the effects of *Panax ginseng* on astrocyte cells under oxidative stress, glutathione reductase (which converts the oxidized glutathione disulfide (GSSG) to its normal reduced state) activity was significantly increased [30]. To our knowledge, only one study has examined the effect of American ginseng on glutathione levels in vivo. In this study, American ginseng did not substantially improve glutathione regeneration in mouse hepatic tissue and also did not show significant in vitro antioxidant activity in the DPPH assay [31]. Further investigation needs to be completed to determine the mechanism by which ginseng exerts antioxidant and anti-diabetic effects.

7.4 Conclusions and Future Directions

This is the first attempt to examine family-held ginseng landraces that have been in cultivation for several decades. Clearly, there is evidence of variation between these landraces. Although HPLC-DAD analysis clearly quantified ginsenoside levels and showed variability, it did not present unique characteristics that would rapidly identify each landrace. Currently, Nuclear Magnetic Resonance (NMR) methods are

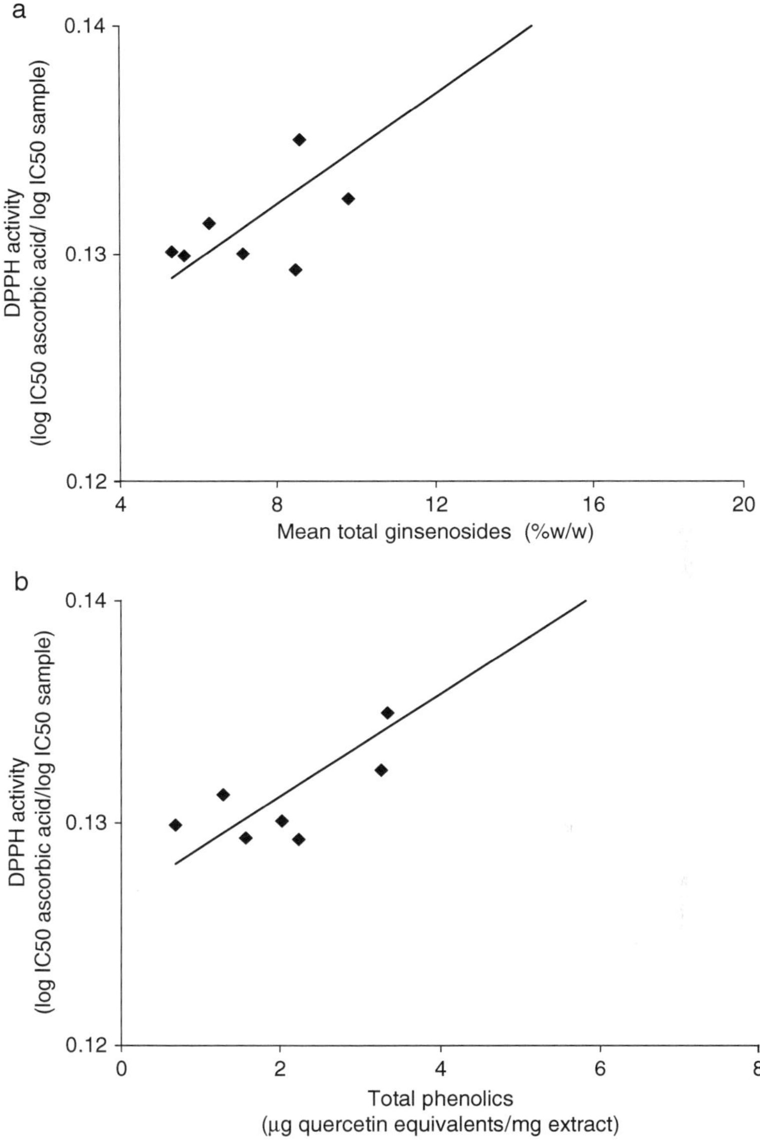

Fig. 7.3 Radical scavenging activity as measured in the DPPH assay versus (**a**) total ginsenoside content (% w/w) and (**b**) total phenolics (μg quercetin per mg extract) of eight ginseng root samples ($n = 3$). The log IC_{50} of each sample was divided by the log IC_{50} of the standard ascorbic acid to serve as a comparison to the positive control

being developed in collaboration with Bruker BioSpin to differentiate ginseng populations. This will allow for rapid quality control and standardization and may complement DNA fingerprinting. This will potentially provide a unique fingerprint for each population. As unique landrace characteristics are identified, it is possible that these may be correlated with important activities. This could then lead to the development of unique, easily identifiable cultivars, each with distinct biological uses.

Although significant in vivo anti-diabetic effects have been reported, this did not translate into strong in vitro anti-glycation and antioxidant activities. There may be stronger in vivo antioxidant and anti-glycation activities, which will be an important area of future investigation.

Acknowledgments We thank Paula Brown (British Columbia Institute of Technology) for developing, validating, and sharing the extraction and HPLC-DAD method. This project was funded by the Ontario Ginseng Innovation Research Centre.

References

1. Mihalov JJ, Marderosian AD, Pierce JC, et al (2000) DNA identification of commercial ginseng samples. J Agric Food Chem 48:3744–3752
2. Small E, Catling PM (2003) Canadian medicinal plant biodiversity priority issues. In: Arnason JT (ed) Biodiversity and health: focusing research to policy. NRC Research Press, Ottawa
3. Ontario Ginseng Growers Association (2009) Ginseng Ontario. http://www.ginsengontario.com. Accessed 02 Sep 2009
4. Assinewe VA, Arnason JT, Aubry A, et al (2002) Extractable polysaccharides of *Panax quinquefolius* L. (North American ginseng) root stimulate TNFα production by alveolar macrophages. Phytomedicine 9:398–404
5. Kitts DD, Popovich G, Hu C (2007) Characterizing the mechanism for ginsenoside-induced cytotoxicity in cultured leukemia (THP-1) cells. Can J Physiol Pharmacol 85:1173–1183
6. Dascalu A, Sievenpiper JL, Jenkins AL, et al (2007) Five batches representative of Ontario-grown American ginseng root produce comparable reductions of postprandial glycemia in healthy individuals. Can J Physiol Pharmacol 85:856–864
7. Vuksan V, Starvo MP, Sievenpiper JL, et al (2000) Similar postprandial glycemic reductions with escalation of dose and administration time of American ginseng in type 2 diabetes. Diabetes Care 23:1221–1226
8. Keum YS, Park YY, Lee JM, et al (2000) Antioxidant and anti-tumor promoting activities of the methanol extract of heat-processed ginseng. Cancer Lett 150:41–48
9. Schlag EM, McIntosh MS (2006) Ginsenoside content and variation among and within American ginseng (*Panax quinquefolius* L.) populations. Phytochemistry 67:1510–1519
10. Lim W, Mudge KW, Vermeylen F (2005) Effects of population, age, and cultivation methods on ginsenoside content of wild American ginseng (*Panax quinquefolium*). J Agric Food Chem 53:8498–8505
11. Assinewe VA, Baum BR, Gagnon D, et al (2003) Phytochemistry of wild populations of *Panax quinquefolius* L. (North American ginseng). J Agric Food Chem 51:4549–4553
12. Dong TTX, Cui XM, Song ZH, et al (2003) Chemical assessment of roots of *Panax notoginseng* in China: regional and seasonal variations in its active constituents. J Agric Food Chem 51:4617–4623
13. Bai D, Brandle J, Reeleder R (1997) Genetic diversity in North American ginseng (*Panax quinquefolius* L.) grown in Ontario detected by RAPD analysis. Genome 40:111–115
14. Schulter C, Punja ZK (2002) Genetic variation among natural and cultivated field populations and seed lots of American ginseng (*Panax quinquefolius* L.). Canada Int J Plant Sci 163:427–439
15. Yu R, Brown PN Single Laboratory Validation Study for the Determination of Ginsenoside Content in *Panax ginseng* C.A. Meyer and *Panax quinquefolius* L. Raw Materials and Finished

Products by High Pressure Liquid Chromatography. 5th Annual NHP Research Conference, Toronto, Ontario. March 27-29, 2008

16. Ahmed N (2005) Advanced glycation endproducts – role in pathology of diabetic complications. Diabetes Res Clin Pract 67:3–21
17. Farsi DA, Harris CS, Reid L, et al (2008) Inhibition of non-enzymatic glycation by silk extracts from a Mexican land race and modern inbred lines of maize (*Zea mays*). Phytother Res 22: 108–112
18. McIntyre KL, Harris CS, Saleem A, et al (2009) Seasonal phytochemical variation of anti-glycation principles in lowbush blueberry (*Vaccinium angustifolium*). Planta Med 75:286–292
19. Singh S, Singh RP (2008) In vitro methods of assay of antioxidants: an overview. Food Rev Int 24:392–415
20. Singleton VL, Rossi JA Jr (1964) Colorimetry of total phenolics with phosphomolbdic-phosphotungistic acid reagents. Am J Enol Vitic 16:144–158
21. Harris CS, Lambert J, Saleem A, et al (2008) Antidiabetic activity of extracts from needle, bark, and cone of *Picea glauca*: organ-specific protection from glucose toxicity and glucose deprivation. Pharm Biol 46:126–134
22. Harbilas D, Martineau LC, Harris CS, et al (2009) Evaluation of the antidiabetic potential of selected medicinal plant extracts from the Canadian boreal forest used to treat symptoms of diabetes: Part II. Can J Physiol Pharmacol 87:479–492
23. Li TSC, Mazza G, Cottrell AC, et al (1996) Ginsenosides in roots and leaves of American ginseng. J Agric Food Chem 44:717–720
24. Christensen LP, Jensen M, Kidmose U (2006) Simultaneous determination of ginsenosides and polyacetylenes in American ginseng root (*Panax quinquefolium* L.) by high-performance liquid chromatography. J Agric Food Chem 54:8995–9003
25. Vuksan V, Stavro MP, Sievenpiper JL, et al (2000) American ginseng improves glycemia in individuals with normal glucose tolerance: effect of dose and time escalation. J Am Coll Nutr 19:738–744
26. Vuksan V, Sievenpiper JL, Wong J, et al (2001) American ginseng (*Panax quinquefolius* L.) attenuates postprandial glycemia in a time-dependent but not dose-dependent manner in healthy individuals. Am J Clin Nutr 73:753–758
27. Sievenpiper JL, Arnason JT, Leiter LA, et al (2003) Variable effects of American ginseng: a batch of American ginseng (*Panax quinquefolius* L.) with a depressed ginsenoside profile does not affect postprandial glycemia. Eur J Clin Nutr 57:243–248
28. Kitts DD, Wijewickreme AN, Hu C (2000) Antioxidant properties of a North American ginseng extract. Mol Cell Biochem 203:1–10
29. Cho WCS, Chung WS, Lee SKW, et al (2006) Ginsenoside Re of Panax ginseng possesses significant antioxidant and antihyperlipidemic efficacies in streptozotocin-induced diabetic rats. Eur J Pharmacol 550:173–179
30. Naval MV, Gómez-Serranillos MP, Carretero ME, et al (2007) Neuroprotective effect of a ginseng (*Panax ginseng*) root extract on astrocytes primary culture. J Ethnopharmacol 112:262–270
31. Yim TK, Ko KM, Ko R (2002) Antioxidant and immunomodulatory activities of Chinese tonifying herbs. Pharm Biol 40:329–335

Chapter 8
Heat, Color, and Flavor Compounds in *Capsicum* Fruit

Ivette Guzman, Paul W. Bosland, and Mary A. O'Connell

8.1 Peppers

Peppers, belonging to the *Capsicum* genus, are called by a variety of names including aji, chile, chillie, chili, chile pepper, paprika, and are one of the most popular vegetables and spices around the world. Originating in the New World, peppers are one of the oldest vegetables known. Pre-Columbian cultivation and trade of this plant was extensive throughout Central America as well as portions of North and South America. Archeological evidence shows domesticated pepper starch fossils in southern Mexico 6,000 years ago [1]. Through Portuguese trade ventures, distribution of pepper increased worldwide, resulting in greater cultivation of this crop. They are grown not only to provide a fiery spice but also for their rich pigments used as natural dyes.

Peppers are important and common ingredients in many cuisines; ubiquitous throughout virtually all Asian and Latin American cuisines, this ingredient is also prominently found in many Mediterranean and African dishes. Pepper fruits are processed and sold in two ways: as a fresh product including canned, frozen, and pickled, and as a dehydrated product like cayenne powder. In 2006 the United Nations Food and Agriculture Organization reported that the United States produces 4% of the world's sweet and hot peppers, ranking sixth behind China, Mexico, Turkey, Indonesia, and Spain [2]. According to the USDA, pepper use in the United States increased by 38% from 1996 to 2006. In 2007, per capita use of chile peppers (on a fresh-weight basis) totaled 6.1 pounds. Consumers in the United States now use more peppers than many traditional vegetables, including asparagus (1.3 pounds), cauliflower (2.1 pounds), and green peas (3.3 pounds) [3]. Peppers are one of the fastest growing specialty produce items due primarily to the changing US diet, the search for alternative flavors and coloring agents, and the influence of a diverse immigrant population.

M.A. O'Connell (✉)

Department of Plant and Environmental Sciences, New Mexico State University, Las Cruces, NM 88003, USA

e-mail: moconnel@nmsu.edu

D.R. Gang (ed.), *The Biological Activity of Phytochemicals*, Recent Advances in Phytochemistry 41, DOI 10.1007/978-1-4419-7299-6_8,
© Springer Science+Business Media, LLC 2011

Pepper belongs to the nightshade family, Solanaceae, in the genus *Capsicum*. This family also includes tomato, potato, eggplant, and tobacco. Globally, pepper breeders have developed a large array of pepper varieties ranging in fruit phenotypes depending on heat level, size, shape, and color forms. To date, 25 species have been reported from blocky bell-type pods with no heat to the small pungent piquin type [4]. Within each species there is variation in fruit shape, color, rate of maturation, and degree of hotness. The five species most commonly cultivated are *C. annuum*, *C. baccatum*, *C. chinense*, *C. frutescens*, and *C. pubescens*. The most common method of developing new varieties is inter- and intraspecific hybridization with subsequent backcrosses to generate the desired combinations of characteristics including heat, flavor, and color [5]. As *Capsicum* is a member of the Solanaceae, rapid progress in molecular genetics is possible using the related resources developed in tomato and potato; many regions of the chromosomes of these crops are syntenic [6, 7]. Further, many of the genes identified originally in *Capsicum* as morphological traits have related counterparts in tomato [8, 9].

The three key fruit quality traits, color, heat, and flavor, are independent characteristics. The color of a fruit does not predict the hotness, despite many consumers associating green peppers with no-heat bell peppers and red powder with hot cayenne chile. Hence in New Mexico, the state question is "Red or Green?" to determine which version of the host's salsa or enchilada is hotter. From a phytochemical perspective the independence of these traits relates to the biosynthetic pathways for these compounds and the genes for both the structural and regulatory genes for those pathways. This review describes some of the recent advances in the analyses of the compounds, enzymes, and genes associated with color and flavor in *Capsicum*.

8.2 Pigments

Peppers are rich in phytochemicals, which give them a diverse range of colors consisting of red, yellow, orange, purple, and brown. The pigments include three classes of phytochemicals: chlorophyll, carotenoids, and anthocyanins. The range of pepper fruit color is due to the accumulation of one or more of these pigments. Pigments are photosynthetic compounds involved in many plant functions. These molecules absorb light at different wavelengths and aid in the conversion of light energy into chemical bond energy during photosynthesis [10]. Pigments act as ultraviolet and visible light protectants in plant tissues and can have additional antioxidant properties [11–13]. Not only do pigments function at a cellular level, they also function as attractants for pollinators and seed dispersal agents [14].

8.2.1 Chlorophyll

Chlorophylls are hydrophobic compounds made of four modified pyrole rings with an Mg atom and an attached long C20 hydrocarbon tail (Fig. 8.1). In pepper, unripe fruit colors can vary from ivory, green, or yellow. The green color is accumulation of chlorophyll in the chloroplast while ivory indicates chlorophyll degradation as

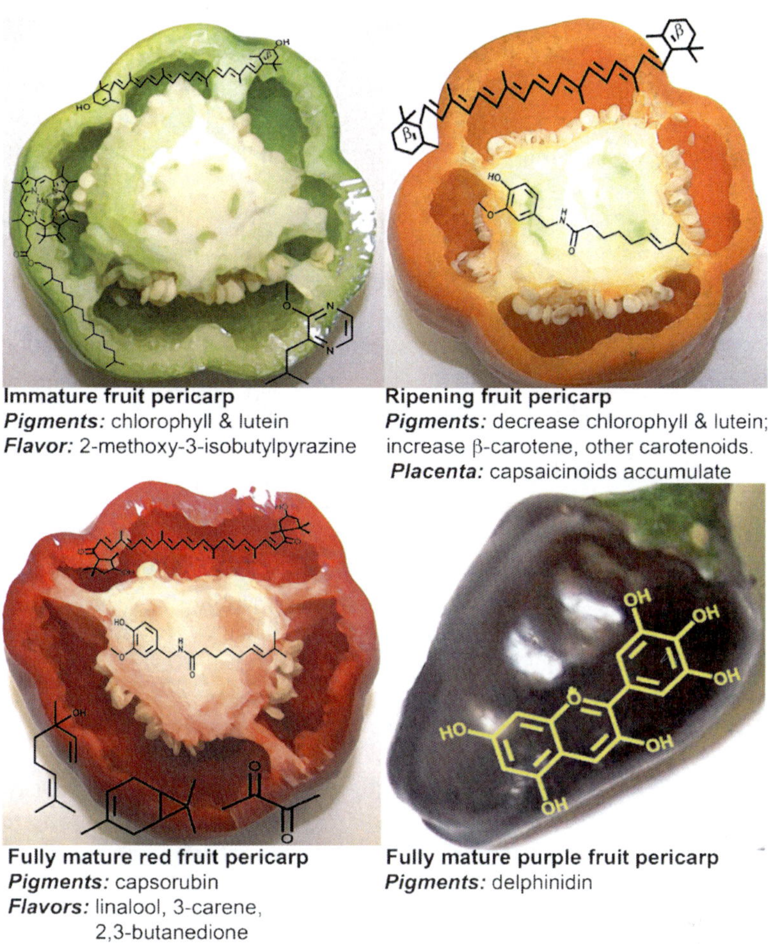

Immature fruit pericarp
Pigments: chlorophyll & lutein
Flavor: 2-methoxy-3-isobutylpyrazine

Ripening fruit pericarp
Pigments: decrease chlorophyll & lutein; increase β-carotene, other carotenoids.
Placenta: capsaicinoids accumulate

Fully mature red fruit pericarp
Pigments: capsorubin
Flavors: linalool, 3-carene, 2,3-butanedione
Placenta: capsaicinoids

Fully mature purple fruit pericarp
Pigments: delphinidin

Fig. 8.1 Four pepper-ripening stages illustrating the structure of the phytochemical in the tissue where it is synthesized

the fruit ripens [15]. There is a condition in peppers however that inhibits the degradation of chlorophyll and is associated with the chlorophyll retainer gene (*cl*) [16]. The persistent presence of chlorophyll in fruit ripening to accumulate other pigments like carotenoids or anthocyanins produces brown or black mature fruit colors (Fig. 8.1). Chlorophyll in black pepper fruit is 14-fold higher compared to violet fruit [17].

8.2.2 Carotenoids

Yellow, orange, and red color in mature pepper fruits is a result of carotenoid metabolism and accumulation in the chromoplasts of the pericarp (Fig. 8.1).

Carotenoids are the largest group of natural plant pigments [10]. Carotenoids help in photosynthesis as light-harvesting pigments given that they absorb at different wavelengths than chlorophyll [18]. They also protect the photosynthetic centers from photooxidative damage by acting as quenchers of triplet excited states in chlorophyll molecules. In 1998, Matsufuji et al. [19] reported that the carotenoid, capsanthin, in peppers had a longer lasting radical scavenging effect compared to other antioxidants. Some peppers can accumulate as much as 1.2 mg total carotenoids per gram fruit fresh weight [20].

Carotenoid catabolic products also have a physiological role in the plant. Oxidative cleavage of carotenoids by carotenoid cleavage dioxygenases (CCDs) generates apocarotenoids [21]. Apocarotenoids serve the plant as antifungal agents or in the synthesis of flavor or aroma of flowers and fruits. A well-known downstream product of an apocarotenoid is abscisic acid (ABA), a phytohormone in plants [21].

During fruit ripening, the pepper carotenoid biosynthetic pathway is able to make a variety of carotenoids including β-carotene, violaxanthin, capsorubin, and capsanthin (Fig. 8.2) [18]. The latter two are ketoxanthophylls specific to peppers and are not found in other plants. Capsanthin and capsorubin accumulate in red-colored peppers, while β-carotene is responsible for orange-colored fruit (Fig. 8.1).

Carotene-rich peppers are important sources of vitamin A, with some varieties accumulating 136 μg of β-carotene per gram fresh weight [20, 22]. Carotenoids are so important that the USDA recommends incorporating orange fruits or vegetables twice a week in a diet [23]. Red and orange peppers, especially, have a high content of carotenoids.

Pepper breeders are interested in making carotenoid-rich peppers not only for higher vitamin A content but also for use in red pigment extraction. The red pigments are extracted as an oleoresin and used as a natural colorant. This dye is used in a variety of products ranging from cosmetics to food coloring [24].

Carotenoids are lipid-soluble terpenoids derived from the isoprenoid pathway and are located in hydrophobic areas of cells. All have a 40-carbon isoprene backbone with a variety of ring structures at one or both ends (Fig. 8.2) [25]. The carbon skeleton is derived from five-carbon isoprenoid groups and contains alternating conjugated double bonds. There are two kinds of carotenoids (Fig. 8.2): carotenes composed of carbon and hydrogen and xanthophylls composed of carbon, hydrogen, and oxygen.

8.3 Biosynthesis of Carotenoids in Peppers

In pepper as in many plants, there are two sources of isoprene monomers: the mevalonic acid pathway and the plastidal pool from pyruvate and glyceraldehyde-3-phosphate [26]. Pepper carotenoid biosynthesis uses the plastidal pathway for the isopentyl pyrophosphate monomers and the resident terpenoid synthases and transferases [27]. Using the 5-carbon isoprene pool, the prenyl transferases sequentially synthesize 10-carbon (geranyl), 15-carbon (farnesyl), and 20-carbon (geranylgeranyl) pyrophosphate polymers (GGPP). The enzyme phytoene synthase, PSY, commits GGPP to the carotenoid pathway. PSY is responsible for the head-to-head condensation of two 20-carbon GGPP molecules to form phytoene. Phytoene, a 40-carbon molecule with alternating carbon double bonds, is the backbone for all carotenoids (Fig. 8.2) [28]. The gene for PSY was one of the first genes on this pathway to be cloned and characterized in pepper [29].

A series of desaturation reactions convert phytoene to ζ-carotene and then to lycopene, the important red pigment in tomatoes. In pepper, lycopene undergoes a cyclization reaction on both ends by lycopene β-cyclase, thus producing β-carotene (Fig. 8.2) [25]. Beta-carotene is then converted to β-cryptoxanthin, zeaxanthin,

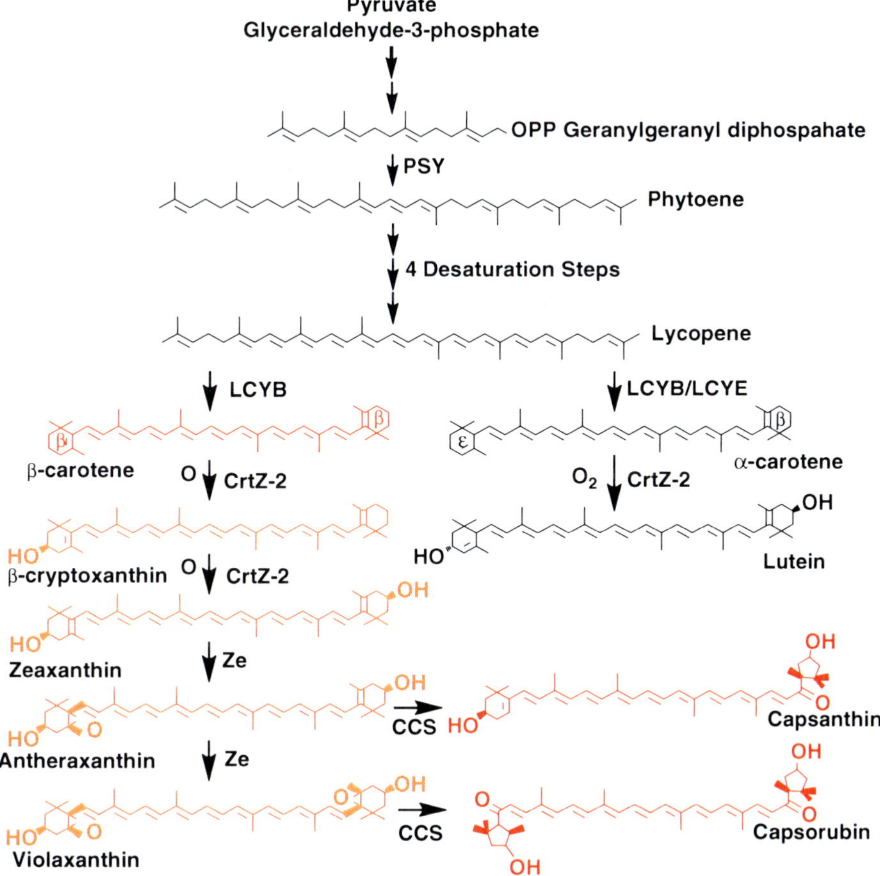

Fig. 8.2 Carotene and xanthophyll biosynthetic pathways in *Capsicum*. Isopentenyl pyrophosphate (IPP); phytoene synthase (PSY); lycopene β-cyclase (LCYB); lycopene ε-cyclase (LCYE); β-carotene hydroxylase (CrtZ-2); zeaxanthin epoxidase (Ze); and capsanthin–capsorubin synthase (CCS)

by β-carotene hydroxylase (*CrtZ-2*) [30]. Zeaxanthin epoxidase generates antheraxanthin and violaxanthin [31]. Capsanthin is made with the aid of an enzyme called capsanthin–capsorubin synthase, CCS [32], which can also convert violaxanthin to capsorubin, the other red pigment in pepper. The desaturases and cyclases that synthesize the specific carotenoids are thought to form a multi-enzyme carotenogenic complex in the thylakoid membrane and stroma of the chromoplasts and chloroplasts [28].

Hurtado-Hernandez and Smith [33] demonstrated that there are three independent loci *C1*, *C2*, and *Y* determining fruit color and eight phenotypes, using an F_2 population of a hybridization between white and red peppers. In this hybridization, red color is dominant, and the F_2 progeny displayed fruit colors ranging from red, orange, peach, and white. Lefebvre et al. [34] demonstrated that *Ccs* is a candidate gene for the *Y* locus, while Huh et al. [35] proposed that phytoene synthase, *Psy*, may be encoded at the *C2* locus. The biochemical function at the *C1* locus, however, has not been determined.

There are, however, two theories as to what makes the orange color in pepper. The orange phenotype could be due to low amounts of red pigments, capsanthin/capsorubin, or no red pigments and high β-carotene concentrations [36]. An inactivation of CCS would result in no synthesis of the red pigments, capsanthin and capsorubin, as seen in several studies [34, 37, 38]. Further, PSY, the rate-limiting enzyme in carotenoid production, may be responsible for the *C2* locus discrimination between red and orange cultivars [35, 39]. Hence, the orange or light colors could be a result of reduced PSY activity, yielding a reduced production of all carotenoids.

8.4 Carotenoid Analysis

The extraction, separation, identification, and quantification methods of carotenoids in peppers have improved over the years. As the fruit ripens xanthophylls become esterified, with 50–80% esterified at one to four possible hydroxyl positions on the rings [40]. Further, this complex mixture is light sensitive and prone to oxidation. Commercial standards are available for the free carotene and xanthophyll structures. Saponification of the carotenoid extract can generate free xanthophyll forms and reduce the complexity of the mixture. Co-migration of the standards and the saponified extract will also allow identification and quantification of carotenoids. Separation by thin-layer chromatography is an excellent tool for rapid and qualitative analyses. Quantification has been carried out using high-performance liquid chromatography (HPLC) [20, 36]. Resolution of these complex mixtures is best achieved with the use of an ultra-performance liquid chromatography (UPLC). Along with better separation, the system uses much smaller volumes; thus, smaller samples are used, sensitivity is increased, and the run times are shorter [36]. The resolution of *Capsicum* pericarp carotenoids by UPLC is demonstrated in Fig. 8.3. Six carotenoid standards were well resolved with base line separation, allowing the

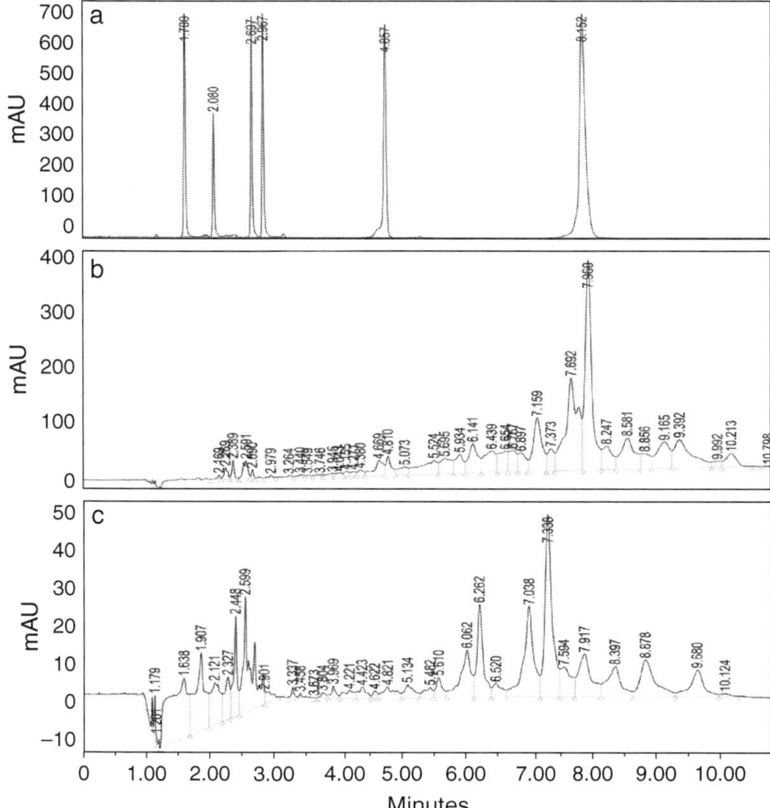

Fig. 8.3 UPLC analysis of *Capsicum*-lyophilized pericarp carotenoids. Carotenoids detected by absorption at 454 nm, following separation on a waters acquity C18 1.8 μm HSS particle, 2.1 × 100 mm column resolved with 10% isopropanol (v/v) (**a**) and 100% acetonitrile (**b**). The solvent profile included two linear phases (0–3 min at 75% (**b**); 3–11 min from 95 to 100%); flow rate of 0.75 mL/min. (**a**) Standards (each at 100 ppm): capsorubin (1.7 min), capsanthin (2.08 min), antherxanthin (2.69 min), zeaxanthin (2.97 min), β-cryptoxanthin (4.86 min), and β-carotene (8.15 min). (**b**) Valencia pericarp extract. (**c**) NuMex Sunset pericarp extract

identification of these carotenoids in complex extraction mixtures from *Capsicum* fruit. Two examples are provided: *C. annuum* cv. Valencia and *C. annuum* cv. NuMex Sunset. Both of these fruit appear orange; however, only Valencia has detectable levels of β-carotene (Fig. 8.3b, c).

The production of β-carotene-rich crops is an important agricultural topic. The availability of carotenoid biosynthetic genes, as well as new molecular techniques, has opened the doors for metabolic engineering. This has resulted in transgenic plants with improved nutritional quality: the development of "golden rice" [41]. The levels of β-carotene in the endosperm (1 to 2 μg β-carotene/g FW rice endosperm) although higher than wild-type rice, as this is a tissue that normally does not accumulate carotene, provide less than 10% of the daily recommended amount [26]. Manipulation of these pathways for increased provitamin A levels in crops like tomato or pepper may be much more productive. *Capsicum* is able to accumulate a diverse number of carotenoids and, in high concentrations, the expression and regulation of the carotenoid genes in pepper could lead to new metabolic engineering tools.

Owing to the variability in color and the reason for the color, further studies are needed to understand how these genes are expressed and how the pathway is regulated in plants of worldwide agronomic importance. Chile pepper ripening is an attractive model to study carotenogenesis because it differentiates green chloroplasts from non-green plastids called chromoplasts, which accumulate large amounts of carotenoids. In particular, xanthophylls are produced in massive quantity, which implies that pepper chromoplasts, unlike tomato chromoplasts that accumulate mainly lycopene, possess enzymes for the whole carotenoid biosynthetic pathway. Molecular markers for carotenoid biosynthetic enzymes have been mapped and are available for *Capsicum annuum* [6, 42].

8.4.1 Anthocyanins

In some peppers, immature pepper fruit and leaves can appear violet to black in color (Fig. 8.1) [17]. These colors are attributed to anthocyanin accumulation in the vacuoles. Anthocyanins are flavonoids synthesized from the phenylpropanoid pathway [43]. The first committed step in the anthocyanin pathway is the synthesis of tetrahydroxychalcone by the enzyme chalcone synthase. These products are then isomerized to make the colorless flavanones, which are further modified to produce the colorful anthocyanins. Black or violet peppers metabolize and accumulate the anthocyanin delphinidin as both an aglycone and a glycosylated compound [17]. The intense black pigmentation in pepper leaves and fruit is characteristic of high concentrations of delphinidin, chlorophyll, and carotenoids, with the leaves accumulating sevenfold higher levels of delphinidin than the fruit. Other than delphinidin, there are no other anthocyanins known to accumulate in peppers. There are at least two loci affecting anthocyanin accumulation in pepper, *A* and a second locus, modifier of *A* (*MoA*), which when expressed, modifies the intensity of the purple color in the presence of A [43]. Ben Chaim et al. [44] mapped *A* to chromosome 10. The gene product at *A* is a myb transcription factor, similar to a gene in petunia that also controls the anthocyanin biosynthetic pathway [45].

8.5 Capsaicinoids

The nitrogenous compounds produced in pepper fruit, which cause a burning sensation, are called capsaicinoids. Capsaicinoids are purported to have antimicrobial effects for food preservation [46], and their most medically relevant use is as an analgesic [47, 48]. Capsaicinoids have been used successfully to treat a wide range of painful conditions including arthritis, cluster headaches, and neuropathic pain. The analgesic action of the capsaicinoids has been described as dose dependent, and specific for polymodal nociceptors. The gene for the capsaicinoid receptor has been cloned (TRPV1) and the receptor transduces multiple pain-producing stimuli [49, 50].

Capsaicinoids all share a common aromatic moiety, the vanillylamine, and differ in the length and degree of unsaturation of the fatty acid side chain (Fig. 8.4) [51, 52]. The perception of burn from these individual capsaicinoids will also vary slightly; capsaicin and dihydrocapsaicin are the hottest and deliver their bite everywhere from the mid-tongue and palate to down in the throat [53]. The traditional method for quantifying the heat of chile preparations has been the organoleptic test designed by Scoville in 1912 [24]. The dilution factor at which heat can no longer be detected is recorded as Scoville Heat Units (SHU). Currently, HPLC analysis of the individual capsaicinoids in organic extractions of chile preparations is the accepted method [54]; however, the unit of description is still commonly SHU, where 1 ppm is equal to 16 SHU.

Capsaicinoids are very potent, humans can perceive >1 ppm; yet capsaicinoids are odorless and tasteless. As birds are insensitive to capsaicinoids, the accumulation

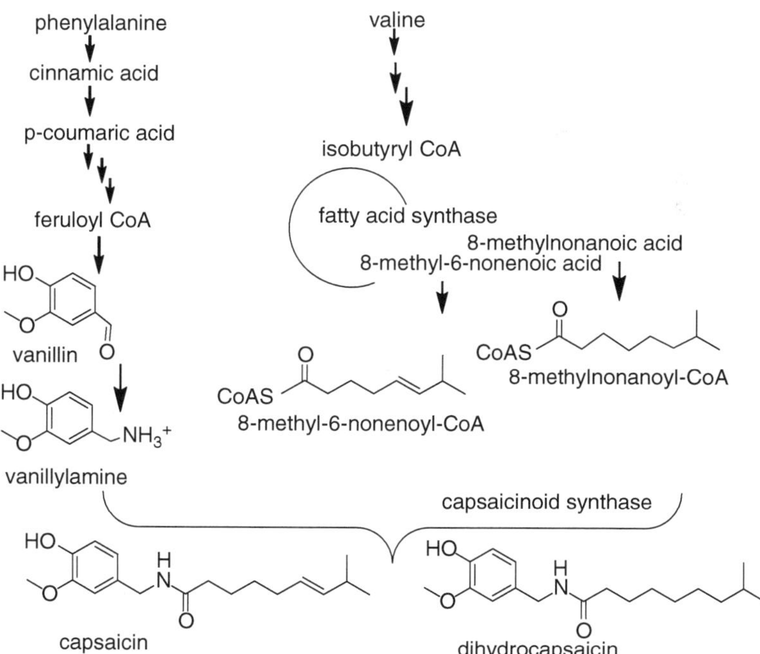

Fig. 8.4 Schema of capsaicinoid biosynthesis

of capsaicinoids in chile serves as herbivory repellents for mammals without interfering with the consumption of the fruit by birds, which serve as seed dispersal agents [55, 56].

8.5.1 Capsaicinoids Are Products of the Phenylpropanoid Pathway and the Branched Chain Fatty Acid Pathway

Capsaicinoids are synthesized by the condensation of vanillylamine with a short chain branched fatty acyl CoA. A schematic of this pathway is presented in Fig. 8.4. Evidence to support this pathway includes radiotracer studies, determination of enzyme activities, and the abundance of intermediates as a function of fruit development [51, 52, 57–63]. Differential expression approaches have been used to isolate cDNA forms of biosynthetic genes [64–66]. As this approach worked to corroborate several steps on the pathway, Mazourek et al. [67] used *Arabidopsis* sequences to design primers to clone the missing steps from a cDNA library. They have expanded the schema to include the biosynthesis of the key precursors phenylalanine and leucine, valine and isoleucine. Prior to this study it was not clear how the vanillin was produced, and thus the identification of candidate transcripts on the lignin pathway for the conversion of coumarate to feruloyl-CoA and the subsequent conversion to vanillin provide key tools to further test this proposed pathway.

Validation of the role of feruloyl-CoA in the synthesis of the vanillin precursor will be detection of the appropriate intermediates and/or enzyme activities in placental extracts that could account for the production of the predicted levels of capsaicinoids. The presence of low levels of monolignol intermediates could be explained by lignin biosynthesis. An alternate route from phenylalanine to vanillin has been considered by some investigators; Orlova et al. [68] demonstrated the role of the benzenoid pathway in petunia flowers for the biosynthesis of phenylpropanoid/benzenoid volatiles.

The intermediates for the branched chain fatty acid production have been detected in tissues [69]. The saturated and desaturated forms of the branched chain acyl-ACP and acyl-CoA are in the same relative amounts as in the final capsaicinoid products, as demonstrated for two different cultivated species, habanero (*C. chinense*) and jalapeño (*C. annuum*). From these results the authors indicate that the desaturation step occurs prior to release from the FAS complex.

The substrates for capsaicinoid synthase were first defined by Fujiwake et al. [70] to be vanillylamine and 8-methyl-6-nonenoyl CoA. The synthesis of dihydrocapsaicin and the other naturally occurring variants are achieved by condensation of vanillylamine with the respective branched chain acyl, for example, 8-methylnonanoyl-CoA. The gene for capsaicinoid synthase has been linked to an acyltransferase (*At3* or *Pun1*) cloned by Kim et al. [66] and mapped to the *C* locus on chromosome 2 [71, 72]. Direct biochemical confirmation of capsaicinoid synthase remains to be established.

8.5.2 Capsaicinoids Accumulate in the Epidermal Cells of the Placenta

Capsaicinoids start to accumulate 20 days post anthesis and synthesis usually persists through fruit development [59, 62, 73]. The site of synthesis and accumulation of the capsaicinoids is the epidermal cells of the placenta in the fruit (Fig. 8.5) [60, 72, 73]. Ultimately, the capsaicinoids are secreted extracellularly into receptacles between the cuticle layer and the epidermal layer of the placenta [74]. These receptacles of accumulated capsaicinoids are macroscopically visible as pale yellow to orange droplets or blisters on the placenta of many chile types (Fig. 8.5).

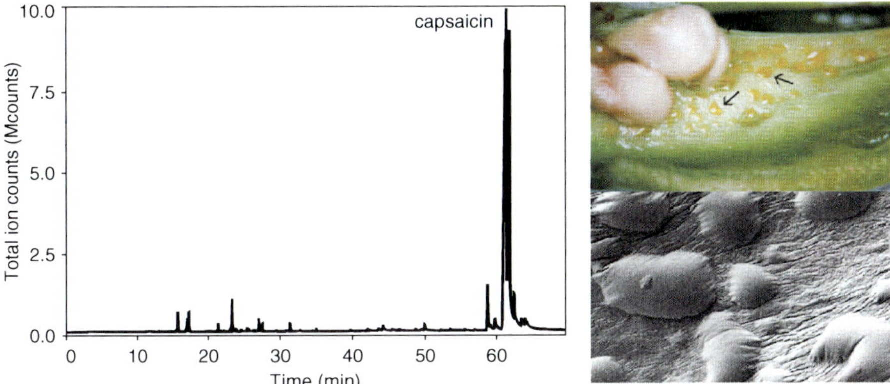

Fig. 8.5 Capsaicin accumulates in blisters/vesicles on surface of placenta. (*left panel*) Total ion chromatogram of oil in habanero vesicle. The oil in the vesicle was collected directly with a Hamilton syringe, diluted with hexane, and analyzed on a varian GC-MS, DB-5 column. The capsaicin peak was identified based on match with NIST MS library. (*right upper panel*) A stereoscope view of habanero placenta (5× magnification) and seeds are visible (*upper left panel*). Arrows indicate blisters. (*right lower panel*) A scanning electron micrograph of habanero placenta (40× magnification). A color version of the image is on *line*

8.5.3 Genetic and Environmental Effects on the Expression of Pungency

The ability to synthesize capsaicinoids is controlled by a single dominant gene, *C* [75], or *Pun1*, mapped to chromosome 2 [71, 72, 76, 77]. No-heat types have been identified among many of the domesticated species and fruit types [78, 79]; bell pepper is an example of a no-heat *C. annuum* var. *annuum*. And among almost all of these mutants, the same locus controls the ability to accumulate capsaicinoids, that is, *Pun1*. These inheritance studies along with the gene sequence data support the hypothesis that the gene product at *Pun1* is capsaicinoid synthase. Confounding the presumption of *Pun1* as the gene for capsaicinoid synthase are the observations by Iwai et al. [80, 81] that capsaicinoids could be detected in non-pungent fruit if stored under continuous light, and placental extracts from non-pungent fruit could synthesize capsaicinoids if vanillylamine and isocapric acid are provided. Together, these results raise the possibility that the gene product at *Pun1* is a regulatory gene or a structural gene upstream in either the phenylpropanoid pathway or the branched chain fatty acid pathway and not capsaicinoid synthase.

Screening for no-heat mutants in *Capsicum* collections and populations is a common practice for pepper breeders, as capsaicinoid accumulation is unique to the members of genus. A second locus for control of capsaicinoid production exists in at least two other species, *C. chinense* and *C. chacoense* [82], and the mutation that revealed this second locus is called *loss of vesicle* (*Lov*). The blisters on the placental surface that fill with capsaicinoids are absent in these no-heat mutants. The gene product and map location of this mutation are not known.

In addition to the simple Mendelian control of the ability to produce capsaicinoids at all controlled by *Pun1* and *Lov*, there is also quantitative genetic inheritance, controlling the degree of heat [83, 84], and several QTLs associated with capsaicinoid accumulation have been reported [77, 85]. A review of this approach for linking genetic markers with capsaicinoid biosynthetic genes is presented by Mazourek et al. [67].

Finally, in addition to the qualitative and quantitative genetic control on the expression of pungency, there is very strong environmental influence [24, 86, 87]. Any environmental stress will increase the heat level. For example, elevated night temperatures and water-deficit conditions induce an increased accumulation of capsaicinoids. The environmental effects are strong enough for mild and moderate cultivars to switch ranks in accumulation of capsaicinoids depending on the environmental conditions of production. Perhaps as a result of these environmental effects, there is extensive fruit-by-fruit variability on a single plant for this trait [88, 89].

8.6 Flavors or Volatile Aroma Compounds

Although most consumers appreciate the fieriness of chile, capsaicinoids are not perceived through odor or taste receptors but through the nociceptive pain receptors described earlier. The compounds in chile fruit that create the flavor and aroma are produced in the fruit wall. Buttery et al. [90] generated vacuum steam distilled oil from green bell pepper macerate, with well over 40 peaks on subsequent GC/MS analysis. Of these peaks, the major flavor compound associated with "bell pepper aroma" was 2-methoxy-3-isobutylpyrazine (Fig. 8.1). They also reported several monoterpenoids in abundance, limonene, trans-β-ocimene, and linalool as well as other aliphatic aldehydes and ketones. The flavor composition of dried red bell pepper powder (sweet paprika) extracted with ether identified 44 key peaks by GC/MS [91]. In these dried samples the key compounds were β-ionone and several furanones. The post-harvest processing and the different fruit maturities as well as possible varietal differences are all causes for the different aromatic profiles.

Solid phase microextraction (SPME) is an ideal approach to monitor volatile flavor components. This approach has been used to identify the volatile compounds in the headspace of fresh fruit during maturation [92]. Using SPME fibers and GC/MS, the key flavor components are hexanal, 2-isobutyl-3-methoxypyrazine, 2,3-butanedione, 3-carene, trans-2-hexenal, and linalool (Fig. 8.1). In this study, the principal aroma compounds whose abundance varied during fruit development were specifically identified.

8.7 Concluding Remarks

The quality traits in *Capsicum* fruit are the result of accumulations of diverse phytochemicals to generate cultivar specific differences in heat, color, and flavor. These chemicals are important nutrients and have medically relevant bioactive properties. The production and accumulation of these compounds in their fruit tissues depend greatly on four factors: the genetic background of the variety, developmental regulation of the structural genes, specificity of gene expression depending on organ or tissue, and the impact of environmental factors. For more than 4,000 years, people in the Americas have selected for these traits giving us an immensely diverse germplasm. Today the international community has generated and shared molecular markers and DNA sequence information that allows us to rapidly link genes, enzymes, and metabolite levels to phenotypic characters. *Capsicum* provides a wealth of resources for biochemical geneticists seeking to understand the expression and regulation of genes in networks controlling heat, color, and flavor. Early on in this review we implied that these key quality traits in *Capsicum* fruit were unrelated. From a systems biology perspective that will clearly not prove to be true; if we learn how to model the flow of metabolites from photosynthesis through primary metabolism and then through these diverse organ-specific biosynthetic pathways, we will undoubtedly describe how the production of red carotenoids in fruit pericarp impacts the levels of capsaicinoids in fruit placenta.

Acknowledgment This work was supported in part by NM Agricultural Experiment Station, NIH grants GM08136 and GM61222, and USDA grant CSREES 2009-34604-19939.

References

1. Perry L, Dickau R, Zarillo S, Holst I, Pearsall DM, Piperno DR, Berman MJ, Cooke RG, Rademaker K, Ranere AJ, Ramond JS, Sandweiss DJ, Scaramelli F, Tarble K, Zeidler JA (2007) Starch fossils and the domestication and dispersal of chili peppers (*Capsicum spp. L.*) in the Americas. Science 315:986–988
2. FAO (2006) Food and Agriculture Organization of the United States. http://www.fao.org
3. USDA (2008) Vegetables and Melons Outlook. http://www.ers.usda.gov/publications/vgs/2008/08Aug/VGS328.pdf
4. Pickersgill B (2007) Domestication of plants in the Americas: insights from Mendelian and molecular genetics. Ann Bot 100:925–940

5. Bosland PW (1996) Capsicums: Innovative uses of an ancient crop. In: Janick J (ed) Progress in new crops. ASHS Press, Arlington, VA, 479–487

6. Livingstone KD, Lackney VK, Blauth JR, Van Wijk R, Jahn MM (1999) Genome mapping in capsicum and the evolution of genome structure in the Solanaceae. Genetics 152:1183–1202

7. Wu F, Eannetta NT, Xu Y, Durrett R, Mazourek M, Jahn MM, Tanksley SD (2009) A COSII genetic map of the pepper genome provides a detailed picture of synteny with tomato and new insights into recent chromosome evolution in the genus *Capsicum*. Theor Appl Genet 118:1279–1293

8. Paran I, Van Der Knaap E (2007) Genetic and molecular regulation of fruit and plant domestication traits in tomato and pepper. J Exp Bot 58:3841–3852

9. Wang D, Bosland P (2006) The genes of *Capsicum*. HortScience 41:1169–1187

10. DellaPenna D, Pogson BJ (2006) Vitamin synthesis in plants: tocopherols and carotenoids. Annu Rev Plant Biol 57:711–738

11. Deepa N, Kaur C, George B, Singh B, Kapoor HC (2005) Antioxidant constituents in some sweet pepper (*Capsicum annuum* L.) genotypes during maturity. Food Sci Technol 40: 121–129

12. Gross J (1991) Pigments in vegetables: chlorophylls and carotenoids. Van Nostand Reinhold, New York, NY

13. Sergio AR, Paiva MD, Rusell RM (1999) β-carotene and other carotenoids as antioxidants. J Amer Coll Nutr 18:426–433

14. Schemske DW, Bradshaw HD (1999) Pollinator preference and the evolution of floral traits in monkeyflowers (*Mimulus*). Proc Natl Acad Sci USA 96:11910–11915

15. Wang HC, Huang XM, Hu GB, Yang Z, Huang HB (2005) A comparative study of chlorophyll loss and its related mechanism during fruit maturation in the pericarp of fast- and slow-degreening litchi pericarp. Sci Hortic 106:247–257

16. Smith PG (1950) Inheritance of brown and green mature color in peppers. J Hered 41:138–140

17. Lightbourn GJ, Griesbach RJ, Novotny JA, Clevidence BA, Rao DD, Stommel JR (2008) Effects of anthocyanin and carotenoid combinations on foliage and immature fruit color of *Capsicum annuum* L. J Hered 99:105–111

18. Deli J, Molnar P, Toth G (2001) Carotenoid composition in the fruits of red paprika (*Capsicum annuum* var. *lycopersiciforme rubrum*) during ripening; biosynthesis of carotenoids in red paprika. J Agric Food Chem 49:1517–1523

19. Matsufuji H, Nakamura H, Chino M, Takeda M (1998) Antioxidant activity of capsanthin and the fatty acid esters in paprika (*Capsicum annuum*). J Agric Food Chem 46:3468–3472

20. Wall MM, Waddell CA, Bosland PW (2001) Variation in β-carotene and total carotenoid content in fruits of *Capsicum*. HortScience 36:746–749

21. Auldrige ME, McCarty DR, Klee HJ (2006) Plant carotenoid cleavage oxygenases and their apocarotenoid products. Curr Opin Plant Biol 9:315–321

22. von Lintig J, Vogt K (2004) Vitamin A formation in animals: molecular identification and functional characterization of carotenoid cleaving enzymes. J Nutr 134:251S-256S

23. USDA (2000) Nutrition and Your Health: Dietary Guidelines for Americans. http://www. cnpp.usda.gov/Publications/DietaryGuidelines/2000/2000DGProfessionalBooklet.pdf.

24. Purseglove J, Brown E, Green C, Robbins S (1981) Spices. Longman, London

25. Hugueney P, Badillo A, Chen H-C, Klein A, Hirschberg J, Camara B, Kuntz M (1995) Metabolism of cyclic carotenoids: a model for the alteration of this biosynthetic pathway in *Capsicum annuum* chromoplasts. Plant J 8:417–424

26. Fraser P, Bramley P (2004) The biosynthesis and nutritional uses of carotenoids. Prog Lipid Res 43:228–265

27. Hugueney P, Bouvier F, Badillo A, Quennemet J, d'Harlingue A, Camara B (1996) Developmental and stress regulation of gene expression for plastid and cytosolic isoprenoid pathways in pepper fruits. Plant Physiol 111:619–626

28. Cunningham FX, Grantt E (1998) Genes and enzymes of carotenoid biosynthesis in plants. Annu Rev Plant Physiol Plant Mol Biol 49:557–583

29. Romer S, Hugueney P, Bouvier F, Camara B, Kuntz M (1993) Expression of the genes encoding the early carotenoid biosynthetic enzymes in *Capsicum annuum*. Biochem Biophys Res Commun 196:1414–1421
30. Bouvier F, Keller Y, d'Harlingue A, Camara B (1998) Xanthophyll biosynthesis: molecular and functional characterization of carotenoid hydroxlases from pepper fruits (*Capscium annuum* L.). Biochim Biophys Acta 1391:320–328
31. Bouvier F, d'Harlingue A, Hugueney P, Marin E, Marion-Poll A, Camara B (1996) Xanthophyll biosynthesis: cloning, expression, functional reconstitution, and regulation of β-cyclohexenyl carotenoid epoxidase from pepper (*Capsicum annuum*). J Biol Chem 271:28861–22867
32. Bouvier F, Hugueney P, d'Harlingue A, Kuntz M, Camara B (1994) Xanthophyll biosynthesis in chromoplasts: isolation and molecular cloning of an enzyme catalyzing the conversion of 5,6-epoxycarotenoid into ketocarotenoid. Plant J 6:45–54
33. Hurtado-Hernandez H, Smith PG (1985) Inheritance of mature fruit color in *Capsicum annuum* L. J Hered 76:211–213
34. Lefebvre V, Kuntz M, Camara B, Palloix A (1998) The capsanthin-capsorubin synthase gene: a candidate gene for the *y* locus controlling the red fruit colour in pepper. Plant Mol Biol 36:785–789
35. Huh JH, Kan BC, Nahm SH, Kim S, Ha KS, Lee MH, Kim BD (2001) A candidate gene approach identified phytoene synthase as the locus for mature fruit color in red pepper (*Capsicum* spp.). Theor Appl Genet 102:524–530
36. Guzman I, Hamby S, Romero J, Bosland PW, O'Connell MA (2010) Variability of Carotenoid Biosynthesis in Orange Colored *Capsicum* spp. Plant Science, 179:49–59
37. Lang YQ, Yanagawa S, Sasanuma T, Sasakuma T (2004) Orange fruit color in *Capsicum* due to deletion of capsanthin-capsorubin synthesis gene. Breed Sci 54:33–39
38. Popovsky S, Paran I (2000) Molecular genetics of the y locus in pepper: its relation to capsanthin-capsorubin synthase and to fruit color. Theor Appl Genet 101:86–89
39. Thorup TA, Tanyolac B, Livingstone KD, Popovsky S, Paran I, Jahn M (2000) Candidate gene analysis of organ pigmentation loci in the Solanaceae. Proc Natl Acad Sci USA 97: 11192–11197
40. Hornero-Mendez D, Minguez-Mosquera MI (2000) Xanthophyll esterification accompanying carotenoid overaccumulation in chromoplasts of *Capsicum annuum* ripening fruits is a constitutive process and useful for ripeness index. J Agric Food Chem 48:1617–1622
41. Ye X, Al-Babili S, Kloti A, Zhang J, Lucca P, Beyer P, Potrykus I (2000) Engineering provitamin A (β-carotene) biosynthesis pathway into (carotenoid-free) rice endosperm. Science 287:303–305
42. Paran I, Van Der Voort JR, Lefebvre V, Jahn MM, Landry L, Van Schriek M, Tanyolac B, Caranta C, Chaim AB, Livingstone K, Palloix A, Peleman J (2004) An integrated genetic linkage map of pepper (*Capsicum* spp.). Mol Breed 13:251–261
43. Daskalov S, Poulos JM (1994) Updated Capsicum gene list. Capsicum Eggplant Newsl 13: 15–26
44. Ben Chaim A, Paran I, Grube R, Jahn MM, Van Wijk R, Peleman J (2001) QTL mapping of fruit related traits in pepper (*Capsicum annuum*). Theor Appl Genet 102:1016–1028
45. Borovsky Y, Oren-Shamir M, Ovadia R, De Jong W, Paran I (2004) The *A* locus that controls anthocyanin accumulation in pepper encodes a MYB transcription factor homologous to *Anthocyanin2* of *Petunia*. Theor Appl Genet 109:23–29
46. Billing J, Sherman PW (1998) Antimicrobial functions of spices: why some like it hot. Quart Rev Biol 73:3–49
47. Govindarajan V, Sathyanarayana M (1991) *Capsicum* - production, technology, chemistry, and quality. Part V - Impact on physiology, pharmacology, nutrition and metabolism; structure, pungency, pain and desensitization sequences. Crit Rev Food Sci Nutr 29:435–474
48. Winter J, Bevan S, Campbell EA (1995) Capsaicin and pain mechanisms. Brit J Anaesth 75:157–168

49. Caterina MJ, Schumacher MA, Tominaga M, Rosen TA, Levine JD, Julius D (1997) The capsaicin receptor: a heat-activated ion channel in pain pathway. Nature 389:816–824
50. Tominaga M, Caterina MJ, Malmberg AB, Rosen TA, Gilbert H, Skinner K, Raumann BE, Basbaum AI, Julius D (1998) The cloned capsaicin receptor integrates multiple pain-producing stimuli. Neuron 21:531–543
51. Bennett DJ, Kirby GW (1968) Constitution and biosynthesis of capsaicin. J Chem Soc C: 442–446
52. Leete E, Louden M (1968) Biosynthesis of capsaicin and dihydrocapsaicin in *Capsicum frutescens*. J Am Chem Soc 90:6837–6841
53. Krajewska A, Powers J (1988) Sensory properties of naturally occurring capsaicinoids. J Food Sci 53:902–905
54. Collins MD, Mayer-Wasmund L, Bosland PW (1995) Improved method for quantifying capsaicinoids in Capsicum using high performance liquid chromatography. HortScience 30:137–139
55. Tewksbury JJ, Nabhan GP (2001) Directed deterrence by capsaicin in chillies. Nature 412:403–404
56. Tewksbury J, Reagan K, Machnicki N, Carlo T, Haak D, Calderon-Penaloza A, Levey D (2008) Evolutionary ecology of pungency in wild chilies. Proc Natl Acad Sci USA 105:11808–11811
57. Fujiwake H, Suzuki T, Iwai K (1982) Capsaicinoid formation in the protoplast from the placenta of *Capsicum* fruits. Agric Biol Chem 46:2591–2592
58. Fujiwake H, Suzuki T, Iwai K (1982) Intracellular distributions of enzymes and intermediates involved in biosynthesis of capsaicin and its analogues in *Capsicum* fruits. Agric Biol Chem 46:2685–2689
59. Hall R, Holden M, Yeoman M (1987) The accumulation of phenylpropanoid and capsaicinoid compounds in cell cultures and whole fruit of the chili pepper, *Capsicum frutescens* Mill. Plant Cell Tissue Organ Cult 8:163–176
60. Iwai K, Suzuki T, Fujiwake H (1979) Formation and accumulation of pungent principle of hot pepper fruits, capsaicin and its analogues in *Capsicum annuum* var. *annuum* cv. Karayatsubusa at different growth stages after flowering. Agric Biol Chem 43:2493–2498
61. Ochoa-Alejo N, Gomez-Peralta JE (1993) Activity of enzymes involved in capsaicin biosynthesis in callus tissue and fruits of chili pepper (*Capsicum annuum* L.). J Plant Physiol 141:147–152
62. Sukrasno N, Yeoman MM (1993) Phenylpropanoid metabolism during growth and development of *Capsicum frutescens* fruits. Phytochemistry 32:839–844
63. Suzuki T, Kawada T, Iwai K (1981) The precursors affecting the composition of capsaicin and its analogues in the fruits of *Capsicum annuum* var. *annuum* cv. Karayatsubusa. Agric Biol Chem 45:535–537
64. Aluru MR, Mazourek M, Landry LG, Curry J, Jahn MM, O'Connell MA (2003) Differential expression of fatty acid synthase genes, *Acl, Fat* and *Kas* in *Capsicum* fruit. J Exp Bot 54:1655–1664
65. Curry J, Aluru M, Mendoza M, Nevarez J, Melendrez M, O'Connell MA (1999) Transcripts for capsaicinoid biosynthetic genes are differentially accumulated in pungent and non-pungent *Capsicum* spp. Plant Sci 148:47–57
66. Kim M, Kim S, Kim S, Kim B-D (2001) Isolation of cDNA clones differentially accumulated in the placenta of pungent pepper by suppression subtractive hybridization. Mol Cells 11: 213–219
67. Mazourek M, Pujar A, Borovsky Y, Paran I, Mueller L, Jahn MM (2009) A dynamic interface for capsaicinoid systems biology. Plant Physiol 150:1806–1821
68. Orlova I, Marshall-Colon A, Schnepp J, Wood B, Varbanova M, Fridman E, Blakeslee JJ, Peer WA, Murphy AS, Rhodes D, Pichersky E, Dudareva N (2006) Reduction of benzoid synthesis in petunia flowers reveals multiple pathways to benzoic acid and enhancement in auxin transport. Plant Cell 18:3458–3475

69. Thiele R, Mueller-Seitz E, Petz M (2008) Chili pepper fruits: presumed precursors of fatty acids characteristic for capsaicinoids. J Agric Food Chem 56:4219–4224
70. Fujiwake H, Suzuki T, Oka S, Iwai K (1980) Enzymatic formation of capsaicinoid from vanillylamine and iso-type fatty acids by cell-free extracts of *Capsicum annuum* var. *annuum* cv. Karayatsubusa. Agric Biol Chem 44:2907–2912
71. Stewart C, Kang B-C, Liu K, Mazourek M, Moore SL, Yoo EY, Kim B-D, Paran I, Jahn MM (2005) The *Pun1* gene for pungency in pepper encodes a putative acyltransferase. Plant J 42:675–688
72. Stewart C, Mazourek M, Stellari GM, O'Connell MA, Jahn M (2007) Genetic control of pungency in *C. chinense* via the *Pun 1* locus. J Exp Bot 58:979–991
73. Suzuki T, Fujiwake H, Iwai K (1980) Intracellular localization of capsaicin and its analogues, capsaicinoid, in *Capsicum* fruit. 1. Microscopic investigation of the structure of the placenta of *Capsicum annuum* var. *annuum* cv. Karayatsubusa. Plant Cell Physiol 21:839–853
74. Ohta Y (1963) Physiological and genetical studies on the pungency of *Capsicum* IV. Secretory organ, receptacles and distribution of capsaicin in the *Capsicum* fruits. Jap J Breed 12: 179–183
75. Heiser CB, Smith PG (1953) The cultivated *Capsicum* peppers. Econ Bot 7:214–227
76. Blum E, Liu K, Mazourek M, Yoo EY, Jahn MM, Paran I (2002) Molecular mapping of the *C* locus for presence of pungency in *Capsicum*. Genome 45:702–705
77. Blum E, Mazourek M, O'Connell MA, Curry J, Thorup T, Liu K, Jahn MM, Paran I (2003) Molecular mapping of capsaciniod biosynthesis genes and QTL analysis for capsaicinoid content in *Capsicum*. Theor Appl Genet 108:79–86
78. Eshbaugh WH (1980) The taxonomy of the genus *Capsicum* (Solanaceae). Phytologia 47:153–166
79. Tewksbury J, Manchego C, Haak D, Levey D (2006) Where did the chili get its spice? Biogeography of capsiaicinoid production in ancestral wild chili species. J Chem Ecol 32:547–564
80. Iwai K, Lee K-R, Kobashi M, Suzuki T (1977) Formation of pungent principles in fruits of sweet pepper, *Capsicum annuum* L. var. *grossum* during post-harvest ripening under continuous light. Agric Biol Chem 41:1873–1876
81. Iwai K, Suzuki T, Lee K-R, Kobashi M, Oka S (1977) In vivo and in vitro formation of dihydrocapsaicin in sweet pepper fruits, *Capsicum annuum* L. var. *grossum*. Agric Biol Chem 41:1877–1882
82. Votava E, Bosland P (2002) Novel sources of non-pungency in *Capsicum* species. Capsicum Eggplant Newslett 21:66–68
83. Garces-Claver A, Gil-Ortega R, Alvarez-Fernandez A, Arnedo-Andres M (2007) Inheritance of capsaicin and dihydrocapsaicin, determined by HPLC-ESI/MS, in an intraspecific cross of *Capsicum annuum* L. J Agric Food Chem 55:6951–6957
84. Zewdie Y, Bosland PW (2000) Capsaicinoid inheritance in an interspecific hybridization of *Capsicum annuum* x *C. chinense*. HortScience 125:448–453
85. Ben Chaim A, Borovsky Y, Falise M, Mazourek M, Kung BC, Paran I, Jahn MM (2006) QTL analysis for capsaicinoid content in *Capsicum*. Theor Appl Genet 113:1481–1490
86. Harvell K, Bosland PW (1997) The environment produces a significant effect on pungency of chiles. HortScience 32:1292
87. Zewdie Y, Bosland PW (2000) Evaluation of genotype, environment, and genotype-by-environment interaction for capsaicinoids in *Capsicum annuum* L. Euphytica 111:185–190
88. Mueller-Seitz E, Hiepler C, Petz M (2008) Chili pepper fruits: content and pattern of capsaicinoids in single fruits of different ages. J Agric Food Chem 56:12114–12121
89. Zewdie Y, Bosland PW (2000) Pungency of chile (*Capsicum annuum* L.) fruit is affected by node position. HortScience 35:1174
90. Buttery RG, Seifert RM, Guadagni DG, Ling LC (1969) Characterization of some volatile constituents of bell peppers. J Agric Food Chem 17:1322–1327

91. Zimmermann M, Schieberle P (2000) Important odorants of sweet bell pepper powder (*Capsicum annuum* cv. *annuum*): differences between samples of Hungarian and Moroccan origin. Eur Food Res Technol 211:175–180
92. Mazidaa MM, Sallehb MM, Osman H (2005) Analysis of volatile aroma compounds of fresh chilli (*Capsicum annuum*) during stages of maturity using solid phase microextraction (SPME). J Food Comp Anal 18:427–437

Chapter 9
Fungal Attack and Cruciferous Defenses: Tricking Plant Pathogens

M. Soledade C. Pedras

9.1 Plant–Pathogen Interactions: A Never-Ending Arms Race

Plants and their pathogens are in close contact through a sophisticated communication system that uses a battery of secondary metabolites aimed at defeating each other's defenses. The communication in such interactions involves a metabolic attack–counterattack that is generally unique to a particular plant–pathogen system. Secondary metabolites with different structures can have equivalent functions in different systems; hence, potential correlations among various systems are initially difficult to make. With time, as the functions of secondary metabolites from plants and pathogens become known, general principles can be envisioned and strategies developed to protect the plant and control the pathogen.

Phytopathogenic fungi are able to infect plants using secondary metabolites with very diverse chemical structures that damage plant tissues irreversibly, thus facilitating attack and colonization. Such phytotoxic metabolites are generally known as phytotoxins, which can be host-selective, if only toxic to plants that host the pathogen, or non-selective when a variety of non-host plants can also be affected. However, it continues to be a greatly challenging task to prove which of these fungal metabolites are produced during infection and indeed that such metabolites are involved in compatible plant–fungus interactions. In addition to phytotoxins, phytopathogenic fungi produce many other groups of metabolites mediating plant defense responses. Among those fungal metabolites are the elicitors, which are of great interest as they induce plant defense responses such as cell wall reinforcement, biosyntheses of phytoalexins, and pathogenesis-related proteins [1]. General elicitors are able to trigger defenses both in host and in non-host plants, whereas race-specific elicitors induce defense responses leading to disease resistance in specific hosts. General elicitors are also called microbial or pathogen-associated molecular patterns (MAMPs or PAMPs) and include lipids, lipopolysaccharides,

M.S.C. Pedras (✉)
Department of Chemistry, University of Saskatchewan, Saskatoon, SK S7N 5C9, Canada
e-mail: s.pedras@usask.ca

D.R. Gang (ed.), *The Biological Activity of Phytochemicals*, Recent Advances
in Phytochemistry 41, DOI 10.1007/978-1-4419-7299-6_9,
© Springer Science+Business Media, LLC 2011

chitin, β-glucans, and fungal-specific glycosylated proteins as well as some smaller molecules [2, 3]. PAMPs sensing by plant transmembrane pattern recognition receptors leads to non-host resistance [4].

To protect themselves from fungal invaders, plants produce physical and chemical/biochemical barriers. In addition, plants have, like animals, some sort of innate immunity, which in plants is due to "pattern recognition receptors" that recognize PAMPs/MAMPs [5]. Once plants detect a pathogen's signal(s), complex mixtures of macromolecules (e.g., callose, chitinases, glucanases, and proteases) and secondary metabolites are produced to deter the invader [6, 7]. Plant defensive metabolites include phytoalexins and phytoanticipins while phytoalexins [8] are biosynthesized de novo in response to biotic and abiotic stresses (e.g., microbial attack, heavy metal salts, or UV radiation), phytoanticipins are constitutive plant defenses whose concentrations can increase under stress conditions [9]. Blends of these metabolites have synergistic activities and additional functions relative to single metabolites, which adds another degree of complexity to the interaction [10].

The "arms race" between plants and their fungal invaders continues to cause enormous crop losses. Despite the widespread use of fungicides, a frequent outcome of the plant–pathogen warfare is the worldwide loss of many primary crops. Indeed, since the 1950s, fungal plant pathogens have been partly controlled by the use of general fungicides; however, because these practices are not sustainable, the discovery of novel strategies that prevent crop losses due to pests and diseases is of vital importance. The ongoing arms race must be controlled; can plants be protected by tricking fungi?

9.2 Pathogen Attack: Phytotoxins and Elicitors from Cruciferous Fungi

Fungal pathogens of crucifers (family Brassicaceae) are very diverse and can have either very high to low economic significance according to the various agricultural regions around the world. Because rapeseed (*Brassica napus* L. and *B. rapa* L.) and canola (*B. napus* L., *B. rapa* L., and *B. juncea* L.) are cultivated worldwide and are of enormous economic significance, their microbial pathogens are among the most investigated. Currently, Phoma blackleg and Alternaria black spot are the most significant diseases of *Brassica* species from a global perspective. The fungal species *Leptosphaeria maculans* (Desm.) Ces. et de Not (asexual stage *Phoma lingam* [Tode ex Fr.] Desm.) and *L. biglobosa* cause blackleg disease (or Phoma stem rot) in rapeseed and canola worldwide [11, 12] including Canada and the United States [13, 14]. *L. maculans* comprises various pathotype groups and subgroups of isolates virulent to global crops (*B. napus*, *B. rapa*, and/or *B. juncea*). Before 2000, all these isolates were included in the species *L. maculans*, but a reclassification introduced the species *L. biglobosa* to enclose the isolates traditionally known as avirulent/weakly virulent to *B. napus* and/or *B. rapa* [15]. However, reverse pathogenicity was observed in two isolates of *L. maculans*, Laird 2 and

Mayfair 2, that is, these isolates were not virulent on canola (*B. napus* and *B. rapa*), but were virulent on brown mustard (*B. juncea*) [16, 17]. Besides attacking important *Brassica* sp., cruciferous condiments such as wasabi (*Wasabia japonica* M.) and white mustard (*Sinapis alba* L.) are also susceptible to blackleg fungi. Alternaria black spot (or dark leaf spot) is also caused by a complex of species that include *Alternaria brassicae* (Berkeley) Saccardo and *A. brassicicola* (Schwein.) Wiltshire. Brown mustard is highly susceptible to black spot disease, which is prevalent in Asia, namely in India. The lack of availability of sources of resistance against Alternaria black spot within Brassicaceae makes Alternaria black spot one of the most damaging and widespread fungal diseases of mustard crops [18].

A recent review of the metabolites of *L. maculans* and *L. biglobosa* produced in diverse culture conditions [19] emphasized that both species biosynthesize host-selective and non-selective phytotoxins. Importantly, it was shown that the composition of metabolite profiles of *L. maculans* depended on the composition of the culture medium. In a chemically defined liquid medium, isolates virulent on canola produced mainly sirodesmin PL (**1**), a non-host-selective phytotoxin, minor sirodesmins with one, three, or four sulfurs bridging the dioxopiperazine ring (sirodesmin H (**3**) [20], sirodesmin J (**4**) and K (**5**) [21]) and phomalirazine (**6**) (Fig. 9.1). The various sirodesmins **1–5** and phomalirazine (**6**) caused necrotic lesions of different intensities on leaves of both resistant and susceptible plants. Phomalide (**7**), the first host-selective phytotoxin isolated from virulent isolates of *L. maculans*, caused disease symptoms (necrotic, chlorotic, and reddish lesions) on canola (susceptible to *L. maculans*) but not on brown mustard or white mustard (species resistant to *L. maculans*) at concentrations ranging from 10^{-5} to 10^{-4} M. Interestingly, it was observed that production of phomalide (**7**) was detected in 30–60-h-old cultures, but once the cultures started to produce sirodesmins the production of phomalide stopped [22]. Phomalide (**7**) was detected in infected leaves of canola, which suggested its involvement in the early stages of plant infection [23]. The detection of phomalide (**7**) but not sirodesmin PL (**1**) in infected leaves

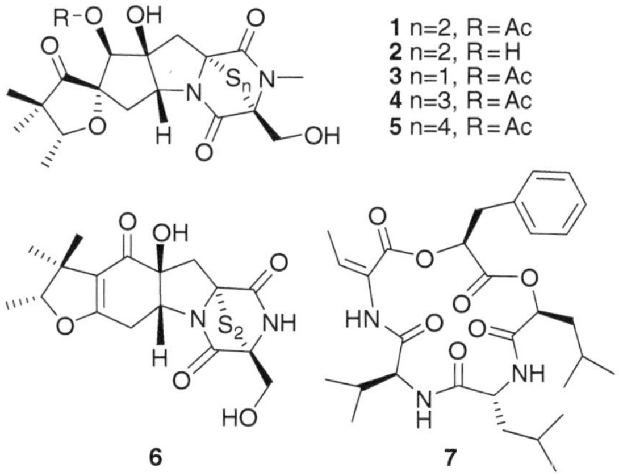

Fig. 9.1 Chemical structures of non-selective phytotoxins **1–6** and host-selective phytotoxin **7** produced by canola virulent isolates of *Leptosphaeria maculans*. Phytotoxins **1–7** are produced in a chemically defined medium

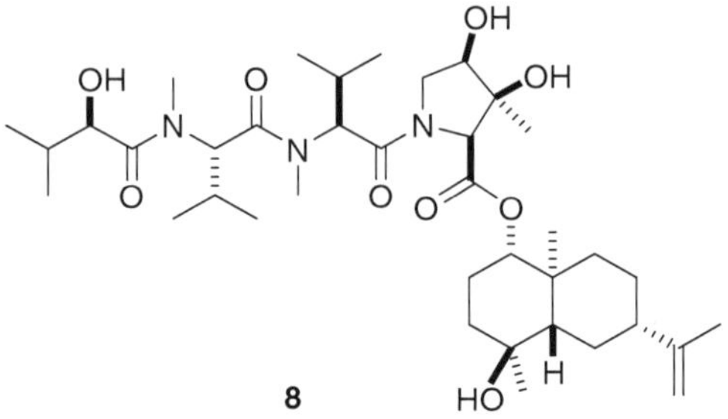

Fig. 9.2 Chemical structure of the host-selective phytotoxin **8** produced by brown mustard virulent isolates of *Leptosphaeria maculans*. Phytotoxin **8** is produced in a chemically defined medium

was consistent with the inhibitory effect of sirodesmin PL (**1**) on the production of phomalide (**7**) in the culture media. Recently however, traces of sirodesmin PL (**1**) were detected in older leaves of canola infected with *L. maculans* using LC-MS [24].

L. maculans isolates Laird 2 and Mayfair 2 (virulent on brown mustard but not on canola) produced in a chemically defined medium the host-selective phytotoxin depsilairdin (**8**) (Fig. 9.2), containing a novel amino acid residue ((2*S*,3*S*,4*S*)-3,4-dihydroxy-3-methylprolyl) and a sesquiterpene moiety (lairdinol A, synthesized recently [25, 26]). Depsilairdin (**8**) caused disease symptoms similar to those caused by the pathogen on brown mustard, that is, strong necrotic and chlorotic lesions, but no lesions on canola.

It was concluded that the sirodesmins biosynthetic genes were not expressed in *L. maculans* cultured in potato dextrose media, where production of maculansin A (**9**), a unique derivative of mannitol, was discovered [27]. It is noteworthy that this phytotoxin was not produced in a chemically defined medium where sirodesmin PL (**1**) was produced in large amounts. Together with maculansin A (**9**), a mixture of epimers of maculansin B (**10**) was isolated from potato dextrose cultures. Maculansin A (**9**) did not show elicitor activity either on resistant or on susceptible plants but was found to cause more tissue damage on plants resistant to *L. maculans/P. lingam* (*B. juncea* L. cv. Cutlass, brown mustard) than on susceptible plants (canola) (Fig. 9.3). This selective reverse phytotoxicity of maculansin A (**9**) continues to be intriguing but appears consistent with reports on expansion of the host range of isolates of *L. maculans* from canola to brown mustard. An additional

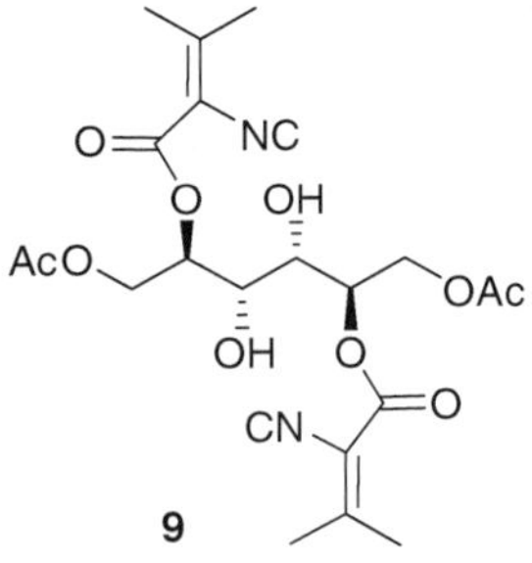

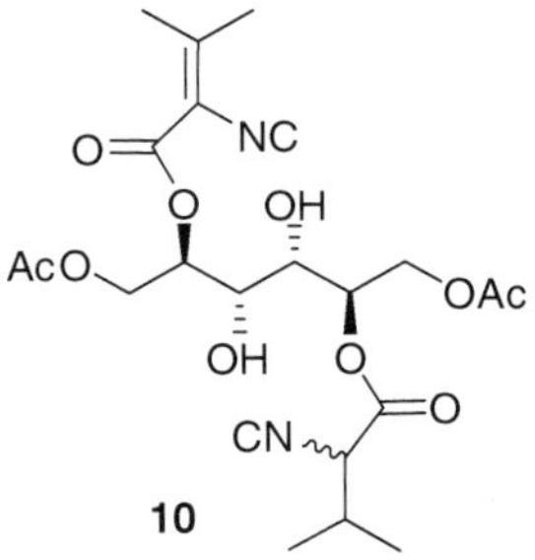

Fig. 9.3 Chemical structures of the selective phytotoxins **9** and **10** produced by canola virulent isolates of *Leptosphaeria maculans*. Phytotoxins **9** and **10** are produced in potato dextrose medium

intriguing aspect in this discovery is the similarity between the chemical structures of maculansin A (**9**) and brassicicolin A (**11**), a host-selective toxin produced by *A. brassicicola* (Fig. 9.3) [28].

The Alternaria black spot fungi *A. brassicae* and *A. brassicicola* produce host-selective toxins as well. While *A. brassicicola* produces brassicicolin A (**11**) as the major host-selective phytotoxin [28], *A. brassicae* produces destruxin B (**12**) (Fig. 9.4) [29]. Consistent with the virulence of these phytopathogens, both brassicicolin A (**11**) and destruxin B (**12**) appeared to be more phytotoxic to the susceptible cruciferous species *B. juncea* than to the tolerant *B. napus*.

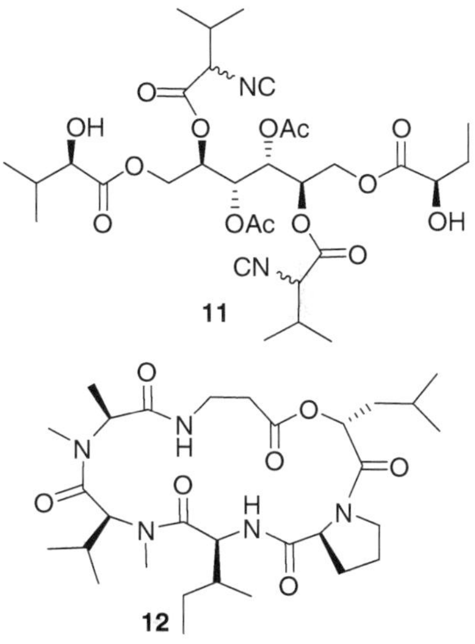

Fig. 9.4 Chemical structures of the selective phytotoxins **11** and **12** produced by mustard virulent isolates of *Alternaria brassicicola* and *A. brassicae*, respectively. Phytotoxins **11** and **12** are produced in a chemically defined medium

Regarding the discovery of novel metabolites from fungal species previously investigated, it is of interest to point out that several fungal genome sequences currently available in public databases have revealed the existence of many "orphan" biosynthetic pathways for which the encoded natural products remain unknown [30]. Therefore, it is likely that further new metabolites and/or phytotoxins will be isolated from fungal species such as *L. maculans* and *A. brassicicola* when grown under different conditions such as culture medium or temperature.

A recent search for general and specific elicitors from *L. maculans* demonstrated that the phytotoxins sirodesmin PL (**1**) and deacetylsirodesmin PL (**2**) are general elicitors since both induced the production of phytoalexins in resistant brown mustard and in susceptible canola [31]. Furthermore, two specific elicitors, a mixture of cerebrosides C (**13**) and D (**14**), were reported from mycelia of liquid cultures of *L. maculans* virulent on canola (Fig. 9.5) [19]. Previously, cerebrosides C (**13**) and D (**14**) were reported from a number of phytopathogenic fungi and were reported to induce the production of phytoalexins in rice plants and disease resistance to the rice blast fungus [32].

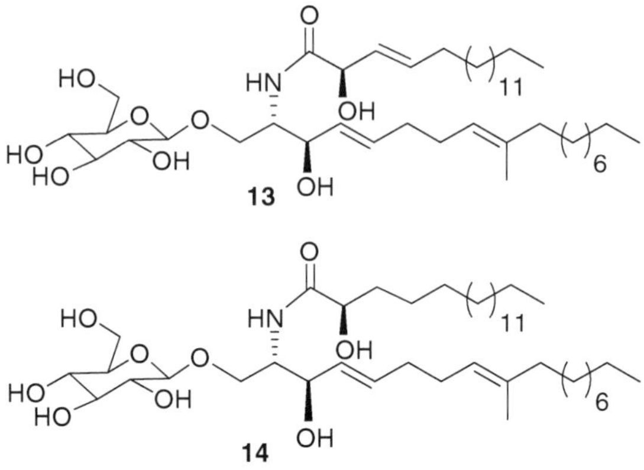

Fig. 9.5 Chemical structures of elicitors from *Leptosphaeria maculans* isolate virulent on canola produced in minimal medium: cerebrosides C (**13**) and D (**14**)

9.3 Cruciferous Defenses: Phytoanticipins and Phytoalexins

The enormous agricultural importance of plants of the family Brassicaceae, commonly known as crucifers, is mainly due to the genus *Brassica*, which includes oilseeds, vegetables, condiments, fodder, and biofuel crops. Several *Brassica* species such as *B. juncea*, *B. napus*, *B. nigra*, *B. oleracea*, and *B. rapa* comprise important subspecies and are cultivated worldwide. For example, *B. rapa* comprises a number of morphologically diverse crops, including the vegetables Chinese cabbage and turnip, and the oilseeds rapeseed, canola, and yellow and brown Sarson. The economic significance of several cultivated *Brassica* species and the amenability to genetic analysis, as well as facile acclimatization explain the enormous amount of research on various species of this family. In addition, a few cruciferous wild species are model systems of great interest due to their smaller genomes than those of *Brassica* species, as for example, *Arabidopsis thaliana* and *Thellungiella* species. *Brassica* species and *A. thaliana* evolved from a common ancestor around 14.5–20.4 million years ago [33, 34], but the haploid genome equivalent of *B. rapa* is about 529 Mb and that of *A. thaliana* is only 125 Mb [35].

Phytoalexins and phytoanticipins are among the cruciferous metabolites that have crucial ecological functions in defense. Wild crucifers of the genus *Thellungiella* appear to show resistance to stress caused by salinity, cold, and draught. The Shandong (from China) and Yukon (from Canada) ecotypes have been cited as *T. halophila*, although both ecotypes appear to belong to the species *T. salsuginea* [36]. The Shandong and Yukon ecotypes are annual crucifers and important model systems due to their small genomes (about twice that of *A. thaliana*) and high resistance to salinity. Both the Shandong and Yukon ecotypes abiotically stressed with $CuCl_2$ produced the phytoalexins 1-methoxybrassenin B (**15**), rapalexin A (**16**), and wasalexins A (**17**) and B (**18**) [37]. The Shandong ecotype exposed to UV radiation (λ_{max} 254 nm) produced the largest quantities of wasalexins A (**17**) and B (**18**), together with the unique wasalexin photoaddition products biswasalexins A1 (**19**) and A2 (**20**) [38]. The production of biswasalexins A1 (**19**) and A2 (**20**) in

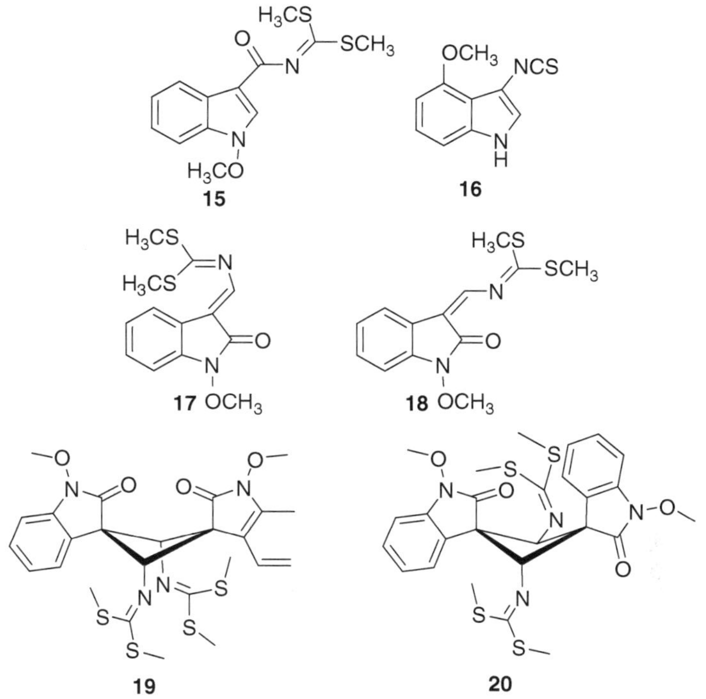

Fig. 9.6 Chemical structures of phytoalexins from *Thellungiella salsuginea* Shandong ecotype: 1-methoxybrassenin B (**15**), rapalexin A (**16**), wasalexins A (**17**) and B (**18**), biswasalexins A1 (**19**), and A2 (**20**)

leaves of UV-stressed Shandong (60 and 15 nmol/g fresh wt, respectively, 2 days after UV elicitation) was thought to result from a photochemical reaction that might protect plants from UV radiation (Fig. 9.6). In addition, perhaps for the first time, it was reported that irrigation with a NaCl solution induced the biosynthesis of phytoalexins, although in smaller amounts than those produced after irradiation of *T. salsuginea* Shandong with UV light. The phytoanticipins indolyl-3-acetonitrile (**21**), arvelexin (**22**), caulilexin C (**23**), neoascorbigen (**24**), and methylthiopropy-lisothiocyanate (**25**) were produced in elicited leaves in much larger amounts than in control leaves [38]. Work is in progress to find additional polar metabolites in the Shandong ecotype (Fig. 9.7).

In general, the phytoalexins of any given plant family are formed from a common primary biosynthetic precursor, but phytoanticipins are usually biosynthesized from rather different precursors. Indeed, most cruciferous phytoalexins are biosynthesized from L-tryptophan while phytoanticipins such as glucosinolates are derived from a wide variety of amino acids (e.g., leucine and many non-protein amino acids, phenylalanine, tryptophan). Interestingly, the cruciferous phytoanticipins glucobrassicin (**29**) [39] and indolyl-3-acetonitrile (**21**) [40] and the phytoalexins derived from brassinin (**28**, R=H) happen to share their primary precursor, i.e., L tryptophan. The

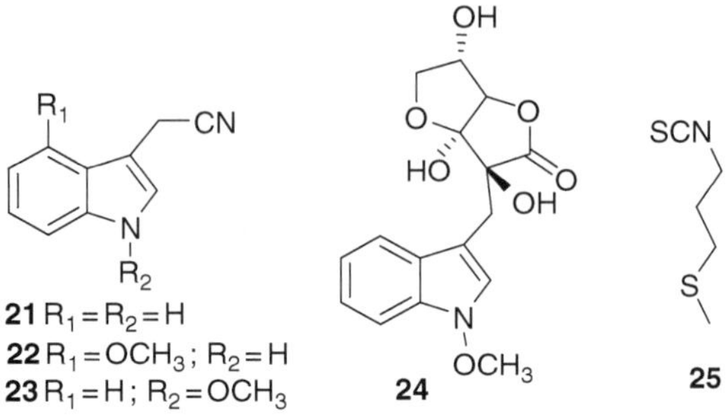

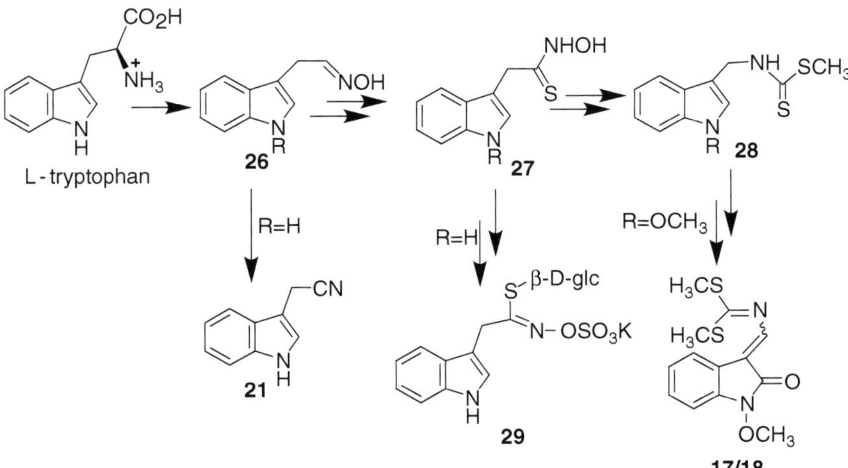

Fig. 9.7 Chemical structures of phytoanticipins from *Thellungiella salsuginea* Shandong ecotype: 3-indolylacetonitrile (**21**), arvelexin (**22**), caulilexin C (**23**), neoascorbigen (**24**), and methylthiopropylisothio-cyanate (**25**)

biosynthetic pathways of brassinin (**28**, R=H) and those of glucobrassicin (**29**) and indolyl-3-acetonitrile (**21**) start with the conversion of L-tryptophan to indolyl-3-acetaldoxime (**26**, R=H) and appear to diverge into various branches [41, 42]. While most of the genes of the glucobrassicin (**29**) biosynthetic pathway [43, 44] and part of the camalexin pathway in *A. thaliana* [45] have been cloned, the brassinin (**28**, R=H) pathway is far from such an achievement. It is obvious that difficulties in tackling this fundamental task are substantial since brassinin (**28**, R=H) does not appear to be produced in *A. thaliana*. However, considering that wasalexins A and B (**17/18**) are likely derived from 1-methoxybrassinin (**28**, R=OCH₃) and that the genome sequencing of *T. salsuginea* is ongoing, this might become a good model plant to identify and clone the genes/enzymes of the brassinin (**28**, R=H) pathway (Fig. 9.8).

Fig. 9.8 Proposed biosynthetic pathway of phytoanticipins **21** and **29** and phytoalexins **17/18** and **28** in *Thellungiella salsuginea* Shandong ecotype (two arrows indicate more than one intermediate in the pathway)

9.4 Pathogen Counterattack: Fungal Detoxifying Enzymes and Paldoxins

Fungal pathogens of crucifers are able to transform phytoanticipins and phytoalexins into harmless products. The outcomes of these reactions, detoxifications, are obviously detrimental to the plant and very useful for the survival of the pathogen. For example, the phytoalexin brassinin (**28**, R=H) is detoxified via oxidation to indole-3-carboxaldehyde (**30**) by *L. maculans* virulent on canola [46], or hydrolysis to indolyl-3-methanamine (**31**) by *L. maculans* [47] and *A. brassicicola* [28], both virulent on brown mustard (Fig. 9.9). Because enzymes that appear to be substrate-specific mediate these detoxifications, inhibition of these enzymes could be a reasonable strategy to deter fungal pathogens. Phytoalexin detoxification inhibitors, coined paldoxins [48, 49], that would allow accumulation of phytoalexins could lead to a selective and environmentally safer control of specific pathogens like *L. maculans* or *A. brassicicola*. Selective inhibitors are less likely to affect non-targeted organisms and thus are anticipated to have a lower negative impact on the cultivated ecosystem.

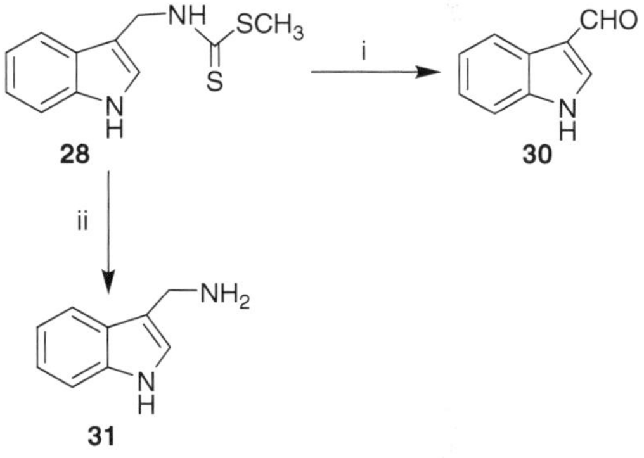

Fig. 9.9 Transformation of the phytoalexin brassinin (**28**) by: (i) *L. maculans* virulent on canola to indole-3-carboxaldehyde (**30**) and (ii) *L. maculans* virulent on brown mustard to indolyl-3-methanamine (**31**)

Brassinin oxidase/dehydrogenase (from *L. maculans* virulent on canola) was purified, characterized, and used to screen libraries of synthetic compounds containing crucifer phytoalexins and analogs of brassinin for potential inhibitory activity [50]. Purified brassinin oxidase/dehydrogenase was ~20% glycosylated, had an apparent molecular mass of 57 kDa, and accepted a wide range of cofactors, from quinones to flavins. A library of brassinin (**28**, R=H) analogs and crucifer phytoalexins was screened for potential inhibition of purified brassinin oxidase/dehydrogenase. Although initially none of the over 70 synthetic analogs of brassinin (**28**, R=H) showed inhibitory activity, it was discovered that the crucifer phytoalexins cyclobrassinin (**32**) and camalexin (**33**) inhibited enzymatic activity competitively [50]. This unexpected result, inhibitory effect of phytoalexins on a fungal detoxifying enzyme, was then reported for the first time [50]. Furthermore, additional synthetic analogs were designed based on the structure of camalexin (**33**) since this phytoalexin was not metabolized by *L. maculans* (Fig. 9.10). Screening

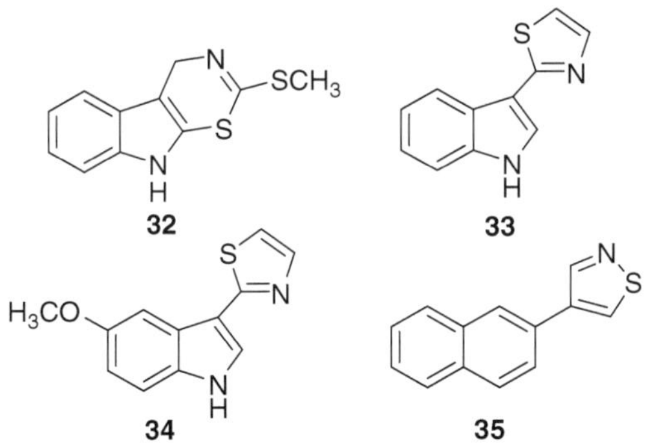

Fig. 9.10 Inhibitors of brassinin oxidase/dehydrogenase, the phytoalexins cyclobrassinin (**32**) and camalexin (**33**) and synthetic inhibitors **34** and **35**

of these camalexin analogs using purified brassinin oxidase/dehydrogenase showed that the inhibitory effect of camalexin derivatives decreased as follows: 5-methoxycamalexin (**34**)>5-fluorocamalexin = 6-methoxycamalexin > camalexin >6-fluorocamalexin [51]. Overall, these results revealed additional biological effects of camalexin (**33**) and its natural derivatives and emphasized that phytoalexins could have positive or negative impacts on plant resistance to different fungal pathogens. Importantly, analogs like **35** displayed lower antifungal activity than brassinin (**28**, R=H) or camalexin (**33**), suggesting that the paldoxin strategy might be a great alternative to control blackleg disease caused by *L. maculans*. Additional lead structures are being used to design paldoxins based on less antifungal scaffolds.

9.5 Concluding Remarks and Future Perspectives

While pathogens of crucifers have mechanisms of attack similar to those of other plant pathogens, many of their phytotoxic metabolites are unique to their own species. The structural variety and number of these fungal metabolites underscore the need to investigate their function in the plant–pathogen interaction. Although this is clearly an area of great importance, progress has been relatively slow as cloning of the biosynthetic genes and generation of corresponding mutants is not a simple task. In this context, the host-selective toxins of both *L. maculans* and *A. brassicicola* are of special interest, the cloning of their genes and generation of mutants would greatly assist in deciphering their roles in the virulence of each pathogen. In addition, the metabolites derived from mannitol, **9–11**, are of particular interest since their biosynthetic pathways are not obvious and no similar mannitol derivatives have been reported from any other organisms.

By definition, cruciferous phytoalexins and phytoanticipins have ecological roles of protecting plants against fungi and other stresses, but the current understanding of such roles and mechanisms of action is still rather incomplete. An important conclusion of the discovery that some cruciferous phytoalexins inhibit brassinin oxidase/dehydrogenase activity stresses that phytoalexins have multiple ecological functions and explains partly some synergistic activities. Application of this discovery led to the development of potent enzyme inhibitors having the camalexin (**33**) scaffold, which in due course may become useful in plant protection against blackleg fungi. In addition, it would be of great importance to generate mutants of *L. maculans* and *A. brassicicola* unable to detoxify brassinin (and other phytoalexins) to evaluate the significance of these brassinin-detoxifying enzymes. Altogether, multiple fundamental and intriguing questions remain regarding crucifers and their interaction with fungal pathogens. If these questions are to be answered within a reasonable time frame, concerted efforts from complementary disciplines including chemistry, biochemistry, and biology are crucial.

Acknowledgments   Support for the author's work reviewed above was obtained from the Natural Sciences and Engineering Research Council of Canada (Discovery Grant), Canada Foundation for Innovation, Canada Research Chairs Program, and the University of Saskatchewan and is gratefully acknowledged.

References

1. Montesano M, Brader G, Palva ET (2003) Pathogen derived elicitors: searching for receptors in plants. Mol Plant Pathol 4:73–79
2. Jones DA Takemoto D (2004) Plant innate immunity: direct and indirect recognition of general and specific pathogen-associated molecules. Curr Opin Immunol 16:48–62
3. Nurnberger T, Brunner F, Kemmerling B, Piater L (2004) Innate immunity in plants and animals: striking similarities and obvious differences. Immunol Rev 198:249–266
4. Jones JDG Dangl JL (2006) The plant immune system. Nature 444:323–329
5. de Wit PJGM (2007) How plants recognize pathogens and defend themselves. Cell Mol Life Sci 64:2726–2732
6. Field B, Jordan F, Osbourn A (2006) First encounters – deployment of defence-related natural products by plants. New Phytol 172:193
7. Maor R Shirasu K (2005) The arms race continues: battle strategies between plants and fungal pathogens. Curr Opin Microbiol 8:399
8. Bailey JA, Mansfield JW (eds) (1982) Phytoalexins. Blackie and Son, Glasgow, UK, p 334
9. VanEtten HD, Mansfield JW, Bailey JA, Farmer EE (1994) Two classes of plant antibiotics: phytoalexins versus "phytoanticipins". Plant Cell 6:1191–1192
10. Pedras MSC, Zheng Q-A, Sarma-Mamillapalle VK (2007) The phytoalexins from Brassicaceae: structure, biological activity, synthesis and biosynthesis. Nat Prod Commun 2:319–330
11. Fitt BDL, Brun H, Barbetti MJ, Rimmer SR (2006) World-wide importance of phoma stem canker (*Leptosphaeria maculans* and *L. biglobosa*) on oilseed rape (*Brassica napus*). Eur J Plant Pathol 114:3–15
12. Howlett BJ, Idnurm A, Pedras MSC (2001) *Leptosphaeria maculans*, the causal agent of blackleg disease of Brassicas. Fungal Gen Biol 33:1–14
13. Chen Y Fernando WGD (2006) Prevalence of pathogenicity groups of *Leptosphaeria maculans* in western Canada and North Dakota, USA. Can J Plant Pathol 28:533–539
14. Kutcher HR, Keri M, McLaren DL, Rimmer SR (2007) Pathogenic variability of *Leptosphaeria maculans* in western Canada. Can J Plant Pathol 29:388–393
15. Shoemaker RA Brun H (2001) The teleomorph of the weakly aggressive segregate of *Leptosphaeria maculans*. Can J Plant Bot 79:412–419
16. Taylor JL, Pedras MSC, Morales VM (1995) Horizontal transfer in the phytopathogenic fungal genus *Leptosphaeria* and host-range expansion. Trends Microbiol 3:202–206

17. Pedras MSC, Taylor JL, Morales VM (1995) Phomaligin A and other yellow pigments in Phoma Lingam and P. Wasabiae. Phytochemistry 38:1215–1222
18. Ghose K, Dey S, Barton H, Loake GJ, Basu D (2008) Differential profiling of selected defence-related genes induced on challenge with *Alternaria brassicicola* in resistant white mustard and their comparative expression pattern in susceptible India mustard. Mol Plant Pathol 9:763–775
19. Pedras MSC Yu Y (2009) Phytotoxins, elicitors and other secondary metabolites from phytopathogenic "blackleg" fungi: structure, phytotoxicity and biosynthesis. Nat Prod Commun 4:1291–1304
20. Pedras MSC, Abrams SR, Séguin-Swartz G (1988) Isolation of the first naturally occurring epimonothiodioxopiperazine, a fungal toxin produced by *Phoma lingam*. Tetrahedron Lett 29:3471–3474
21. Pedras MSC, Séguin-Swartz G, Abrams SR (1990) Minor phytotoxins from the blackleg fungus *Phoma lingam*. Phytochemistry 29:777–782
22. Pedras MSC, Taylor JL, Nakashima TT (1993) A novel chemical signal from the "blackleg" fungus: beyond phytotoxins and phytoalexins. J Org Chem 58:4778–4780
23. Pedras MSC Biesenthal CJ (1998) Production of the host-selective phytotoxin phomalide by isolates of *Leptosphaeria maculans* and its correlation with sirodesmin PL production. Can J Microbiol 44:547–553
24. Elliott CE, Gardiner DM, Thomas G, Cozijnsen A, De Wouw AV, Howlett BJ (2007) Production of the toxin sirodesmin PL by *Leptosphaeria maculans* during infection of *Brassica napus*. Mol Plant Pathol 8:791–802
25. Pedras MSC, Chumala PB, Quail JW (2004) Chemical mediators: the remarkable structure and host-selectivity of depsilairdin, a sesquiterpenic depsipeptide containing a new amino acid. Org Lett 6:4615–4617
26. Pardeshi SG Ward DE (2008) Enantiospecific total synthesis of lairdinol A. J Org Chem 73:1071–1076
27. Pedras MSC Yu Y (2008) Structure and biological activity of maculansin A, a phytotoxin from the phytopathogenic fungus *Leptosphaeria maculans*. Phytochemistry 69:2966–2971
28. Pedras MSC, Chumala PB, Jin W, Islam MS, Hauck DW (2009) The phytopathogenic fungus *Alternaria brassicicola*: phytotoxin production and phytoalexin elicitation. Phytochemistry 70:394–402
29. Gross H (2007) Strategies to unravel the function of orphan biosynthesis pathways: recent examples and future prospects. App Microbiol Biotechnol 75:267–277
30. Pedras MSC, Zaharia IL, Gai Y, Zhou Y, Ward DE (2001) *In planta* Sequential hydroxylation and glycosylation of a fungal phytotoxin: avoiding cell death and overcoming the fungal invader. Proc Natl Acad Sci USA 98:747–752
31. Pedras MSC Yu Y (2008) Stress-driven discovery of metabolites from the phytopathogenic fungus *Leptosphaeria maculans*: structure and activity of leptomaculins A-E. Bioorg Med Chem 16:8063–8071
32. Umemura K, Tanino S, Nagatsuka T, Koga J, Iwata M, Nagashima K, Amemiya Y (2004) Cerebroside elicitor confers resistance to fusarium disease in various plant species. Phytopathology 94:813–818
33. Bowers JE, Chapman BA, Rong J, Paterson AH (2003) Unraveling angiosperm genome evolution by phylogenetic analysis of chromosomal duplication events. Nature 422:433–438
34. Lysak MA Lexer C (2006) Towards the era of comparative evolutionary genomics in Brassicaceae. Plant Syst Evol 259:175–198
35. Johnston JS, Pepper AE, Hall AE, Chen ZJ, Hodnett G, Drabek J, Lopez R, Price HJ (2005) Ev olution of genome size in Brassicaceae. Ann Bot 95:229–235
36. Amtmann A (2009) Learning from evolution: *Thellungiella* generates new knowledge on essential and critical components of abiotic stress tolerance in plants. Mol Plant 2:3–12
37. Pedras MSC Adio AM (2008) Phytoalexins and phytoanticipins from the wild crucifers *Thellungiella halophila* and *Arabidopsis thaliana:* rapalexin A, wasalexins and camalexin. Phytochemistry 69:889–893

38. Pedras MSC, Zheng QA, Shatte G, Adio AM (2009) Photochemical dimerization of wasalexins in UV-irradiated *Thellungiella halophila* and in vitro generates unique cruciferous phytoalexins. Phytochemistry 70:2010–2016
39. Bender J Celenza JL (2009) Indolic glucosinolates at the crossroads of tryptophan metabolism. Phytochem Rev 8:25–37
40. Agerbirk N, Vos M, Kim JH, Jander G (2009) Indole glucosinolate breakdown and its biological effects. Phytochem Rev 8:101–120
41. Pedras MSC Okinyo DPO (2008) Remarkable incorporation of the first sulfur containing indole derivative: another piece in the puzzle of crucifer phytoalexins. Org Biomol Chem 6:51–54
42. Pedras MSC, Okinyo-Owiti DP, Thoms K, Adio AM (2009) The biosynthetic pathway of crucifer phytoalexins and phytoanticipins: de novo incorporation of deuterated tryptophans and quasi-natural compounds. Phytochemistry 70:1129–1138
43. Yan X Chen S (2007) Regulation of plant glucosinolate metabolism. Planta 226:1343–1352
44. Halkier BA Gershenzon J (2006) Biology and biochemistry of glucosinolates. Annu Rev Plant Biol 57:303–333
45. Glawischnig E (2007) Camalexin. Phytochemistry 68:401–406
46. Pedras MSC Taylor JL (1991) Metabolic transformation of the phytoalexin Brassinin by the 'blackleg fungus'. J Org Chem 56:2619–2621
47. Pedras MSC, Gadagi RS, Jha M, Sarma-Mamillapalle VK (2007) Detoxification of the phytoalexin brassinin by isolates of *Leptosphaeria maculans* pathogenic on brown mustard involves an inducible hydrolase. Phytochemistry 68:1572–1578
48. Pedras MSC, Jha M, Ahiahonu PWK (2003) The synthesis and biosynthesis of phytoalexins produced by cruciferous plants. Curr Org Chem 7:1635–1647
49. Pedras, MSC (2005) Protecting plants against fungal diseases: discovering inhibitors of unique metabolic processes occurring in phytopathogenic fungi. Can Chem News 57:16–17
50. Pedras MSC, Minic Z, Jha M (2008) Brassinin oxidase, a fungal detoxifying enzyme to overcome a plant defense: purification, characterization and inhibition. FEBS J 275:3691–3705
51. Pedras MSC, Zoran M, Sarma-Mamillapalle V (2009) Synthetic inhibitors of the fungal detoxifying enzyme Brassinin oxidase based on the phytoalexin camalexin scaffold. J Agric Food Chem 57:2429–2435

Chapter 10
Glucosinolate Degradation Products in Fermented Meadowfoam Seed Meal and Their Herbicidal Activities

Jan F. Stevens and Ralph L. Reed

10.1 Introduction

Meadowfoam is an established oilseed crop primarily grown in the Willamette Valley of western Oregon. The crop plant is derived from *Limnanthes alba* (Hartw. ex Benth., Limnanthaceae), which is endemic to southern Oregon and northern California. The genus *Limnanthes* received agricultural interest in the 1960s as a source for the development of a new oilseed crop [1–3]. In the 1970s, a breeding program with *L. alba* was initiated at Oregon State University with the aim to improve the plant's agronomical properties and oilseed content [4, 5]. The on-going breeding program has resulted in several cultivars that are grown commercially for the cosmetic industry [6–9]. Meadowfoam oil is especially suitable for application in skin care and other cosmetic products due to its high oxidative stability and content of the rare long chain fatty acids, 20:1Δ5, 22:1Δ5, and 22:1Δ13 [10, 11]. The oxidative stability of meadowfoam oil has been attributed to the absence of high levels of oxidation-sensitive polyunsaturated fatty acids and to the presence of antioxidants, such as 1,3-di(3-methoxybenzyl) thiourea [12].

In addition to oil, *Limnanthes* seeds contain high levels of the glucosinolate, glucolimnanthin [13], and minor amounts of rutin [14] and phytoecdysteroids [14–17] (Fig. 10.1). The allelochemical, glucolimnanthin, is readily converted into (phyto)toxic breakdown products when seeds are subjected to physical damage in moist soil. The conversion is mediated by the glucosinolate-degrading enzyme, myrosinase. These conditions are also met when meadowfoam seeds are crushed in the presence of water in the industrial oil extraction process. To prevent myrosinase activity during oil extraction and thus contamination of the oil with lipophilic breakdown products, meadowfoam seeds receive heat treatment to inactivate myrosinase earlier in the extraction process.

J.F. Stevens (✉)

Department of Pharmaceutical Sciences, Oregon State University, Corvallis, OR 97331, USA

e-mail: fred.stevens@oregonstate.edu

D.R. Gang (ed.), *The Biological Activity of Phytochemicals*, Recent Advances in Phytochemistry 41, DOI 10.1007/978-1-4419-7299-6_10,
© Springer Science+Business Media, LLC 2011

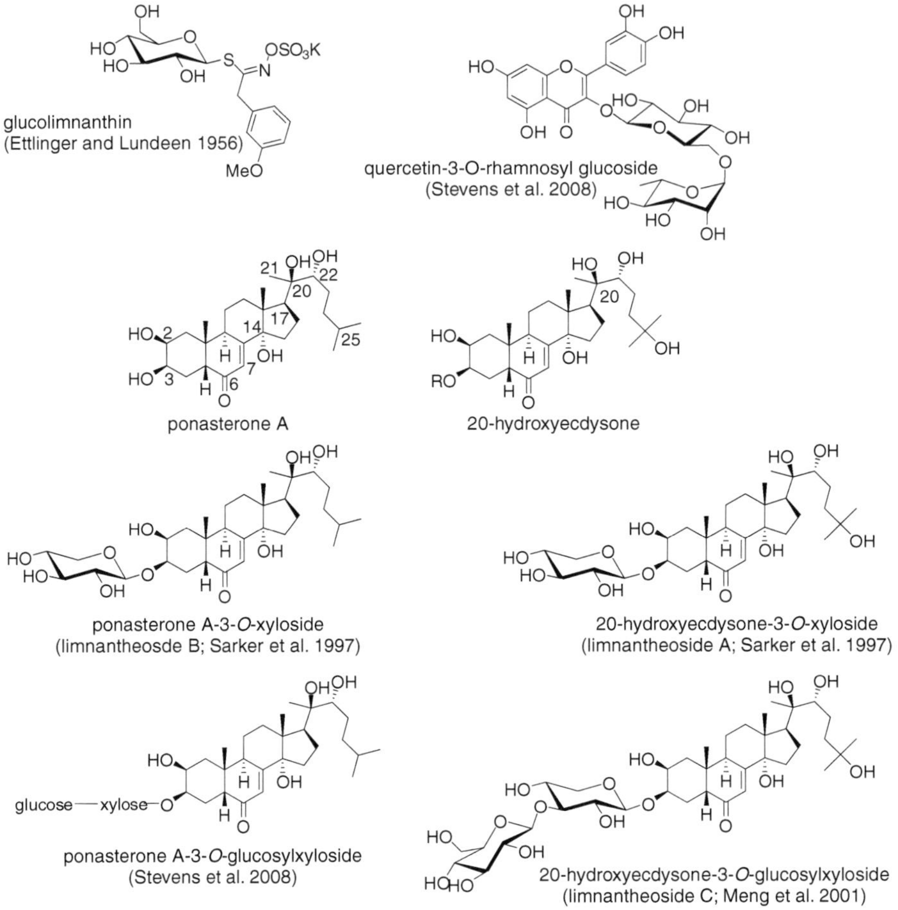

Fig. 10.1 Structures of glucolimnanthin, quercetin-3-*O*-rhamnosylglucoside, and phytoecdysteroids found in *Limnanthes alba* seeds

In comparison with other oilseed crops, meadowfoam is a low-yielding crop and its economic survival can be attributed to its special application in cosmetic products. To increase the economic competitiveness of the crop, commercial uses of the seed meal, a waste product obtained from extracted seeds, are being explored by the meadowfoam industry. Meadowfoam seed meal has received interest as a fertilizer in potatoes and as a pre-emergence herbicide in organic farming. Herbicidal activity of meadowfoam seed meal requires enzyme-mediated conversion of the inactive glucolimnanthin into breakdown products with herbicidal activity. Our research group has explored ways to induce conversion of glucolimnanthin in meadowfoam seed meal, which has no active enzymes due to the above-mentioned heat treatment, by fermenting seed meal with enzyme-active meadowfoam seeds.

Myrosinase has thioglucosidase activity. The myrosinase-mediated degradation of glucolimnanthin yields an intermediate aglycone that rearranges to form

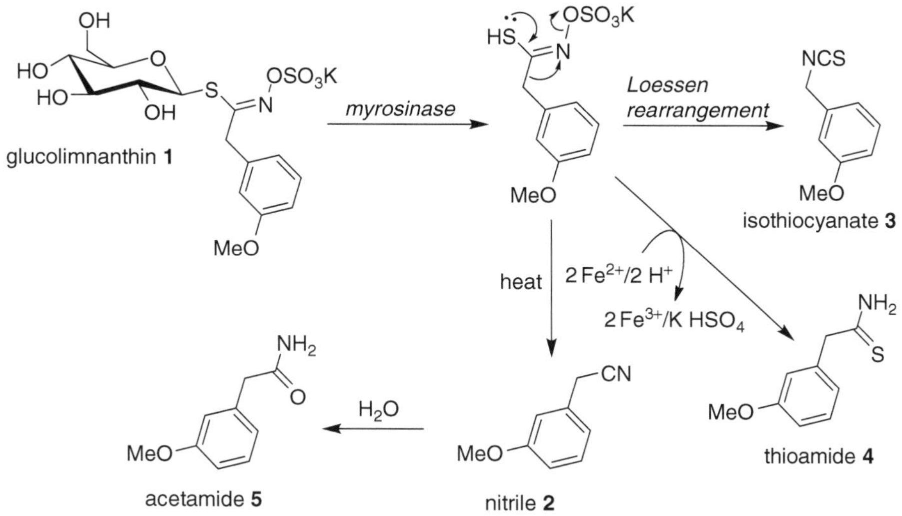

Fig. 10.2 Myrosinase-mediated degradation of glucolimnanthin

3-methoxybenzyl isothiocyanate (**3**) (Fig. 10.2). It is assumed that the rearrangement is also facilitated by myrosinase. The conversion of glucolimnanthin into 3-methoxybenzyl cyanide (nitrile **2**) can be induced by heat but also by Fe^{2+} ions in the presence or absence of myrosinase or other enzymes, such as nitrile-forming proteins [18–21]. Hydrolysis of the nitrile is expected to produce acetamide **5**. The conversion of the intermediate aglycone into thioamide **4** is a reduction reaction and requires reducing equivalents that can by provided by Fe^{2+} ions [22]. We have explored ways to direct glucolimnanthin degradation in meadowfoam seed meal away from isothiocyanate **3** toward nitrile **2** and thioamide **4** with the aim to produce seed meal products with greater herbicidal potency than factory-grade seed meal [23].

10.2 Analytical Procedures

10.2.1 Isolation and Structure Determination of Glucolimnanthin from Meadowfoam Seed Meal

Quantitative analysis of glucosinolates in plant materials requires the availability of pure standards. We isolated gram amounts of glucolimnanthin from meadowfoam seed meal for the purpose of constructing calibration curves and for testing its potency as an herbicide. Most published methods for isolation of glucosinolates involve extraction, ion-exchange chromatography, desalting and purification by size-exclusion chromatography, and finally recrystallization as a tetra-alkyl ammonium salt [24]. We extracted glucolimnanthin from meadowfoam seed meal by soaking in a mixture of water and methanol (1:1, v/v) and percolation with the extraction solvent. The extract was concentrated by rotary evaporation and diluted with methanol. At this point, a white precipitate formed that tested negative for glucolimnanthin by HPLC and that was removed by centrifugation. The supernatant was fractionated on a Sephadex LH-20 column with methanol as the solvent. Fractions containing glucolimnanthin were pooled and purified again by Sephadex LH-20 chromatography using methanol as the solvent. The second chromatographic step and drying *in vacuo* on a lyophilizer yielded >95% pure glucolimnanthin by HPLC and NMR analysis (Fig. 10.3). Further purification by ion-exchange chromatography and recrystallization was therefore not considered necessary. High-resolution electrospray time-of-flight (ToF) mass spectrometry indicated that glucolimnanthin was isolated as the potassium salt: found m/z 478.0269 [M+H]$^+$, calculated for $C_{15}H_{21}NO_{10}S_2K$: 478.0244 (5.2 ppm). This material was used as the standard for HPLC quantification and bioactivity testing. Further details on the isolation and characterization of glucolimnanthin can be found in 23.

10.2.2 HPLC Analysis of Glucolimnanthin

Glucolimnanthin is the predominant glucosinolate in seeds of *L. alba*. Other glucosinolates are found at trace levels in *L. alba* but can be present at significant levels in other species of *Limanthes*. Owing to the presence of the 3-methoxybenzyl moiety in glucolimnanthin, we routinely measure levels of glucolimnanthin in meadowfoam meal and seed samples by high-performance liquid chromatography (HPLC) with UV detection at 274 nm. Many HPLC methods for analysis of glucosinolates have appeared in the literature. Some methods are based on HPLC separation of glucosinolates as their desulfo derivatives on reversed-phase columns [24, 25]. Enzymatic hydrolysis of the sulfate group (e.g., by commercially available aryl sulfate sulfohydrolase from the edible snail *Helix pomatia*) often gives better chromatographic peak shapes and greater resolving power compared to intact glucosinolates [24]. Other HPLC methods employ an ion-pairing agent such as quaternary ammonium salts to improve the chromatographic behavior of intact glucosinolates [24, 26]. The disadvantage of quaternary ammonium salts is that these ion-pairing agents are often incompatible with electrospray liquid chromatography–mass spectrometry (LC–MS). To circumvent this limitation, some authors have used ammonium formate or ammonium acetate to analyze intact glucosinolates by electrospray LC–MS [27]. We have obtained satisfactory results with HPLC-UV and electrospray LC–MS of intact glucolimnanthin and related glucosinolates using a reversed-phase C18 column eluted with a mixture of 0.1% aqueous trifluoroacetic acid (TFA) in acetonitrile (MeCN). Glucolimnanthin is not stable in the presence of 0.1% TFA and therefore we do not add TFA to seed or meal extracts. Acid-mediated degradation of glucolimnanthin does not seem to occur on the HPLC column using gradient elution with 0.1% aqueous TFA and MeCN.

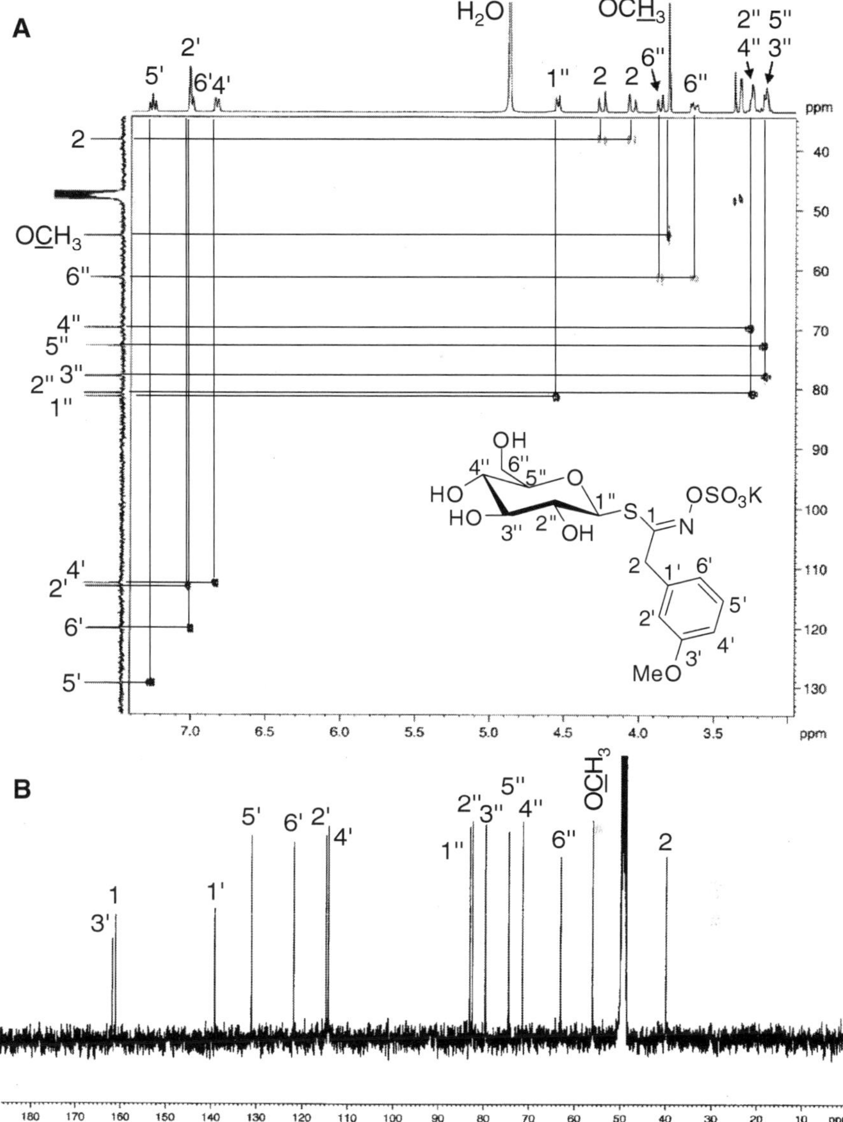

Fig. 10.3 ^{1}H-^{13}C Heteronuclear single quantum coherence (HSQC, *panel* **a**) and ^{13}C NMR (100 MHz) (*panel* **b**) spectral data recorded for glucolimnanthin in CD$_3$OD

Meadowfoam seeds differ from the seed meal in that seeds, even after years of storage, contain active myrosinase whereas the meal does not. In order to prevent myrosinase-mediated degradation of glucolimnanthin during extraction, one could inactivate myrosinase by applying heat or by choosing a solvent in which myrosinase does not exhibit activity. We have found that heating seeds at temperatures of up to 100°C for 30 min does not completely inactivate myrosinase. Above 130°C, no conversion of glucolimnanthin into the corresponding isothiocyanate **3** was observed but we did observe the formation of nitrile **2** and loss of glucolimnanthin. These unsatisfactory findings led us to investigate myrosinase activity in extraction solvents consisting of mixtures of water and MeOH. Formation of isothiocyanate **3** (and nitrile **2**) was prominent in the range 40–80% aqueous MeOH, but minimal at 90% MeOH for both degradation products. The yield of glucolimnanthin was maximal at 90% aqueous MeOH. Therefore we selected 90% aqueous MeOH as the extraction solvent for glucolimnanthin analysis in seeds (Fig. 10.4a).

The yield of glucolimnanthin **1** and nitrile **2** in seed meal by extraction with mixtures of water and MeOH was less critical and maximal at all proportions below 80% MeOH (Fig. 10.4b). For extraction of seed meal we chose 50% aqueous MeOH as the extraction solvent. The extraction yield of glucolimnanthin **1** appears to be poor at 50% aqueous MeOH in Fig. 10.4a, but the low yield in this case is primarily due to enzyme-mediated conversion into its breakdown products **2** and **3**.

Fig. 10.4 Extraction of meadowfoam seeds (*panel* **a**) and seed meal (*panel* **b**) with mixtures of MeOH and water. Method: Meadowfoam seed (cv. Ross) or meadowfoam seed meal was ground in a coffee grinder for 1 min and 0.4 g aliquots were mixed with 8 mL of methanol/water mixtures in screw-capped glass centrifuge tubes, vortexed for 20 s, and sonicated for 60 s. After letting it stand overnight at room temperature in the dark, the samples were vortexed and sonicated again and then centrifuged twice, for 5 min, on a clinical centrifuge, followed by centrifugation of the supernatant for 10 min at 13,000 rpm using a microcentrifuge, and the supernatants were analyzed by HPLC-UV as described in Stevens et al. (2009)

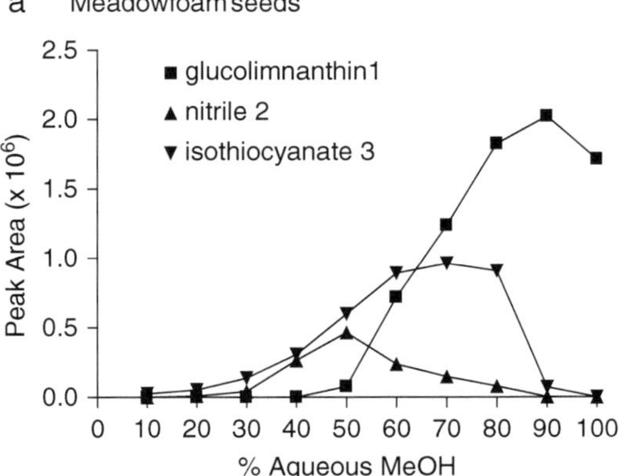

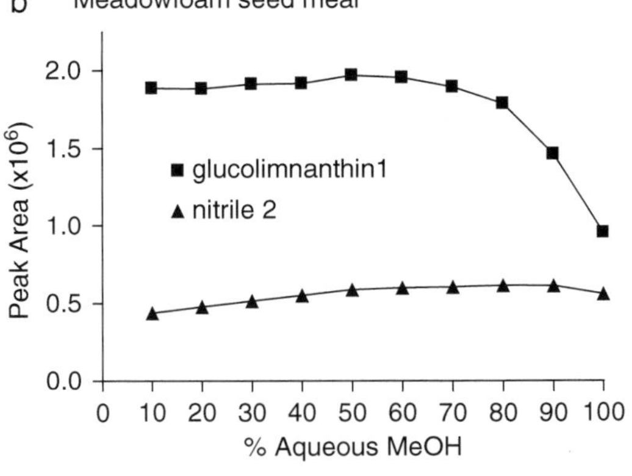

10.2.3 Characterization of Glucosinolates by HPLC–Mass Spectrometry Using Precursor Ion Scanning

LC–MS has been extensively applied to the qualitative and quantitative analysis of glucosinolates, either intact or as their desulfo derivatives [27–31]. Electrospray is the most often used ionization source for detection of intact glucosinolates in the negative ion mode. Popular mass analyzers include three-dimensional ion traps (IT) [28], time-of-flight (ToF) mass analyzers [31], and single (Q) or triple quadrupoles (QqQ) [32, 33], or a combination of these mass analyzers, for example, Q-ToF and QIT mass analyzers [34]. Mass fragmentation of intact glucosinolate anions can be achieved by collision-induced dissociation in most of the hybrid instruments, resulting in the formation of HSO_4^- ions with m/z 97. We and others [35] have exploited this characteristic fragment ion for the detection of glucosinolates of unknown mass on a triple quadrupole mass spectrometer operated in the precursor ion scanning mode. This mode has the unique capability of detecting molecular ions that produce the fragment ion of interest (m/z 97), by scanning the first quadrupole (Q1) within a mass range of expected molecular ions, fragmenting molecular ions in Q2, and setting Q3 at a constant m/z value corresponding to the fragment ion of interest. When precursor ion scanning is applied to the LC–MS/MS analysis of an extract of *L. alba* seed meal, the major chromatographic peak reveals a molecular anion with m/z 438, consistent with the [M-H]$^-$ ion of glucolimnanthin. A seed extract of *L. floccosa* ssp. *pumila* reveals the presence of minor glucosinolates in addition to the major glucolimnanthin (Fig. 10.5). The masses of the minor glucosinolates are consistent with benzyl glucosinolate (glucotropaeolin, m/z 408) and 3-hydroxybenzyl glucosinolate (glucolepigramin, m/z 424) or its 4-hydroxybenzyl isomer, sinalbin (Fig. 10.5).

10.2.4 Identification of the Aglycone Moiety of the L. floccosa Glucosinolate with m/z 424

From a biosynthetic viewpoint it is more likely that the m/z 424 glucosinolate of *L. floccosa* ssp. *pumila* bears the hydroxyl substituent at position 3 (glucolepigramin) than at position 4 (sinalbin) because glucolimnanthin is the *O*-methyl derivative of 3-hydroxybenzyl glucosinolate. Experimental evidence for the 3-hydroxybenzyl moiety was obtained by converting the glucosinolate into its corresponding nitrile by enzyme-mediated glucosinolate degradation in the presence of $FeSO_4$. The source of the hydrolytic enzyme(s) was the seed material itself. Nitrile formation from the glucosinolate of interest was monitored by HPLC with photo-diode array detection and the product chromatographically compared with the commercially available 4-hydroxybenzyl cyanide, 3-hydroxybenzyl cyanide, and 3-methoxybenzyl cyanide. Figure 10.6 shows that the benzyl cyanide product of *L. floccosa* glucosinolate is indeed 3-hydroxybenzyl cyanide and not the 4-hydroxy isomer. The degradation procedure converts glucolimnanthin in extracts of *L. alba* into 3-methoxybenzyl

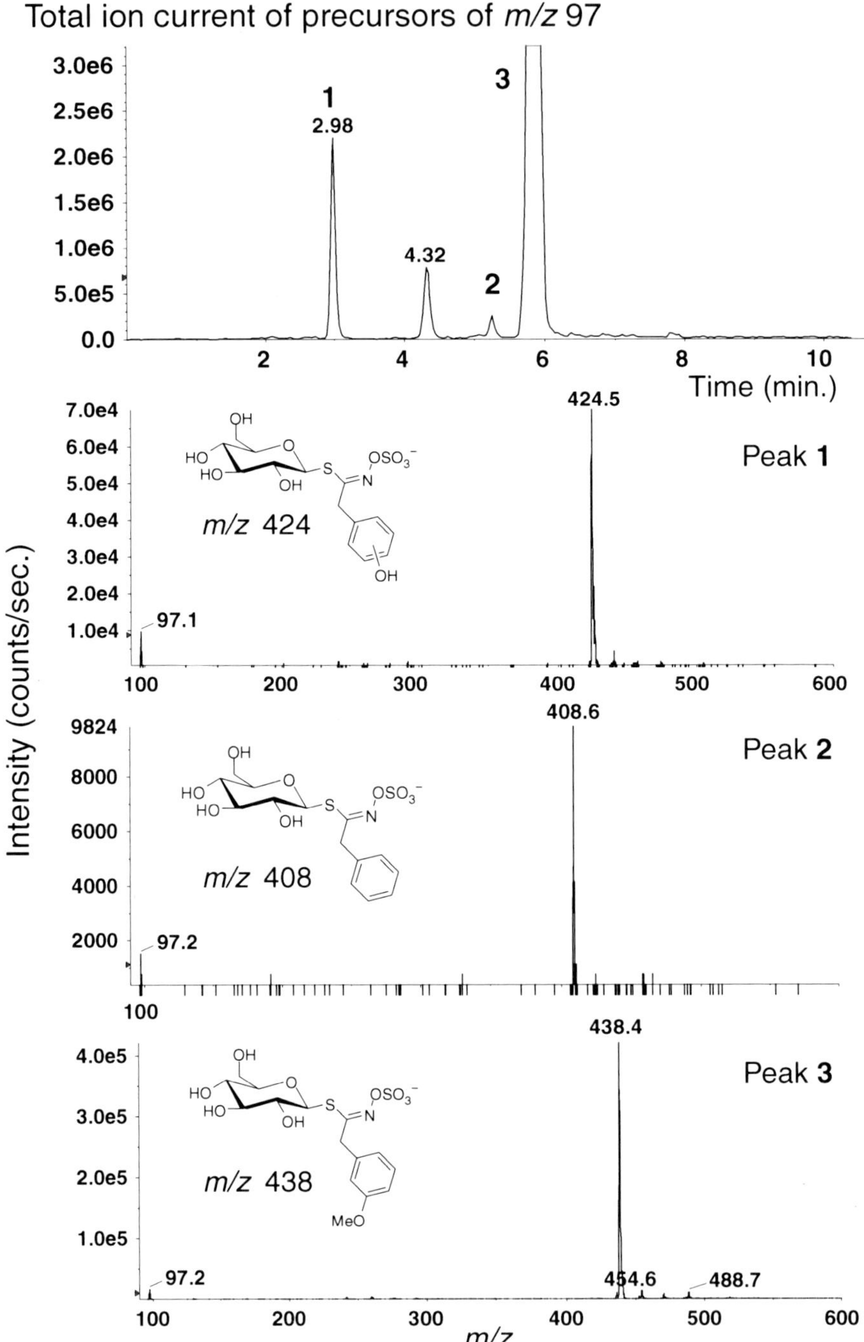

Fig. 10.5 (continued)

Fig. 10.6 HPLC-UV Chromatograms of glucosinolate-derived nitriles in fermented *Limnanthes* seeds supplemented with FeSO$_4$. Method: Three-seed samples and 0.4 mL of an aqueous FeSO$_4$ solution (10 mM) were placed in 1.2 mL tubes (Corning, Corning, NY) with two alloy dowel pins (3/32″ × 5/16″). The seeds were disintegrated using a Retsch MM300 Matrix Mill (Qiagen, Valencia, CA). The extracts were sonicated, allowed to stand overnight at room temperature, and mixed with 0.4 mL MeOH. The tube contents were sonicated again and centrifuged for 10 min at 2,500 g. Supernatants were centrifuged for 10 min at 13,000 rpm and analyzed by HPLC as described in 23

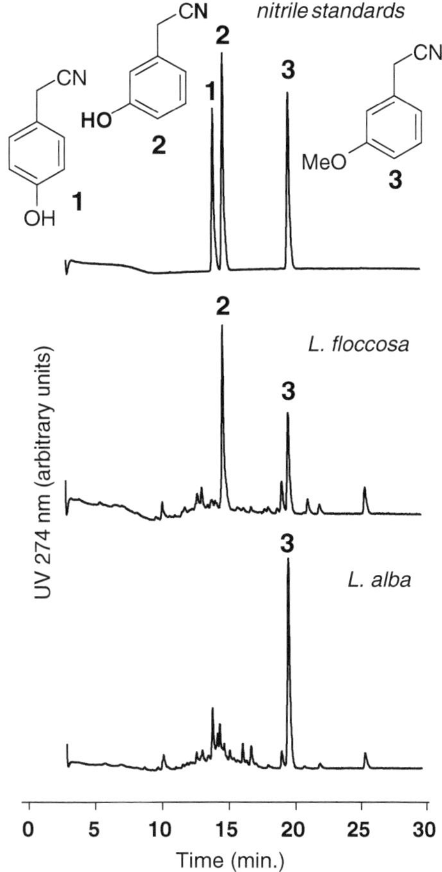

Fig. 10.5 (continued) LC-MS/MS Analysis of an extract of *Limnanthes floccosa* seeds by precursor ion scanning of the HSO$_4^-$ ion with *m/z* 97. Method: Four-seed samples were placed in 2 mL screw-top microtubes (VWR, West Chester, PA) with one mL of 90% methanol and two 4 mm stainless steel balls. The seeds were disintegrated using a GenoGrinder 2000 tissue homogenizer (SPEX CertiPrep, Metuchen, NJ) for 3 min at 1,100 strokes per min, repeated three times, then sonicated for 10 min, allowed to stand for 3 h at room temperature, re-sonicated, and centrifuged for 5 min at 5,500 rpm. The resulting supernatants were centrifuged for 10 min at 13,000 rpm and separated on a 4 × 250 mm Lichrosphere 5 μm C18 column (Phenomenex, Torrance, CA) eluted with a linear gradient of 5–100% acetonitrile and aqueous 0.1% formic acid at 0.8 mL/min. Negative precursor ion scanning of *m/z* 97 was performed using a 3200 QTrap hybrid triple-quadrupole/linear ion trap mass spectrometer (Applied Biosystems/MDS Sciex, Concord, ON, Canada) with a Turbo V electrospray ion source operated at 550°C and –4.5 kV, using a declustering potential of –50 v and a collision energy of –33 eV. The compound with a retention time of 4.32 min could not be identified

cyanide. It was not possible to use the intact glucosinolates or the myrosinase-mediated formation of isothiocyanates as a basis for chromatographic comparison because glucolepigramin and 3-hydroxybenzyl isothiocyanate are not commercially available. Isolation and identification by NMR was not an option either due to limited availability of the seed material from *L. floccosa* plants.

10.3 Herbicidal Activity of Glucolimnanthin and Its Degradation Products

10.3.1 Fermentation of Meadowfoam Seed Meal

The growing interest in sustainable agriculture has led to the search for environment-friendly herbicides. Glucosinolate-producing plants from *Brassica* and *Sinapis* have received much attention as possible sources of natural weed killers [36–38]. Agricultural practices that make use of glucosinolates as pro-herbicides include planting of glucosinolate producers between crop plants and applying freshly cut over-ground parts of glucosinolate producers ("green manure") [39] or glucosinolate-containing seed meals to soil [40]. These approaches rely on the conversion of inactive glucosinolates into active breakdown products by active myrosinase in the leaves or on conversion of glucosinolates in myrosinase-inactivate seed meals by soil microorganisms. Our approach to release active glucosinolate degradation products from myrosinase-inactive meadowfoam meal is based on inoculation of meadowfoam seed meal with a small amount of enzyme-active meadowfoam seeds (1% by weight). 3-Methoxybenzyl cyanide (=3-methoxyphenyl acetonitrile, nitrile **2**) was identified as a phytotoxin present in meadowfoam that inhibited radicle elongation in velvet leaf (*Abutilon theophrasti*) and wheat (*Triticum aestivum*) [41]. Though less abundant than the parent glucolimnanthin, 3-methoxybenzyl cyanide (nitrile **2**) is a component of meadowfoam seed meal that is produced by heat-induced degradation of glucolimnanthin in the industrial oil extraction process (Fig. 10.7). Inoculation of meadowfoam seed meal with 1% of ground meadowfoam seeds in the presence of water caused complete consumption of glucolimnanthin within 3–5 h and concomitant formation of 3-methoxybenzyl isothiocyanate (Fig. 10.8a). The content of nitrile **2** was not affected in this fermentation experiment. To a certain extent, the formation of isothiocyanate **3** can be prevented and the formation of nitrile **2** can be enhanced by a factor of 2–3 by applying a solution of 10 mM $FeSO_4$ instead of water to the seed meal (Fig. 10.8b). In addition, a minor degradation product, thioamide **4**, is also formed in the presence of $FeSO_4$. In both meal fermentation experiments, the levels of nitrile **2** remained constant after consumption of the parent compound **1** whereas the levels of the isothiocyanate **3** declined rapidly at first and more slowly later. The decrease in isothiocyanate **3** is likely due to covalent interaction with proteins in the meal.

The role of ferrous ions in modifying the enzymatic degradation of glucolimnanthin is not completely understood. It is known that species of *Brassica* produce the so-called nitrile-forming enzymes, such as the epispecifier protein, that require ferrous ions at micromolar concentrations [18]. When we incubated

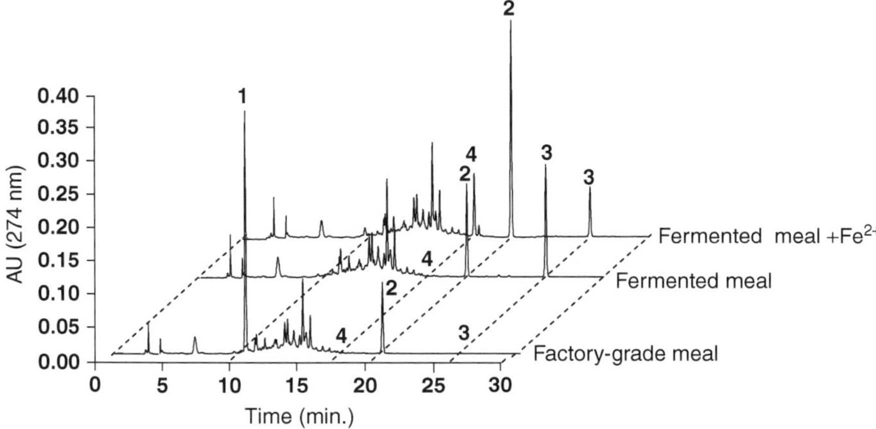

Fig. 10.7 HPLC-UV Analysis of meadowfoam seed meal and fermented meal. Methods: For extraction of meal, 6 ml of 50% aqueous MeOH and meal (1.0 g) were mixed in a screw-capped glass centrifuge tube, vortexed for 30 s, and sonicated for 60 s. After letting it stand overnight at room temperature in the dark, the samples were vortexed and sonicated again and then centrifuged twice, 5 min on a clinical centrifuge, followed by centrifugation of the supernatant for 10 min at 13,000 rpm using a microcentrifuge. The resulting supernatant was diluted 1:9 with 50% methanol. Fermented meals were prepared as described in the caption of Fig. 10.8 (22 h incubation) and extracted as described for untreated seed meal. Sample solutions of 10 μl were analyzed by HPLC using a Lichrosphere 5 μm C18 column (4 × 250 mm, Phenomenex, Torrance, CA) with a linear gradient from 5 to 100% acetonitrile (solvent B) in 0.1% aqueous trifluoroacetic acid (solvent A) over 30 min at 1.0 mL/min. UV absorption was monitored from 210 to 500 nm and peak areas were calculated using 274 nm for all compounds except 270 was used for thioamide **4**

meadowfoam meal with 1% seeds and micromolar concentrations of $FeSO_4$, no nitrile formation was observed, suggesting the absence of active nitrile-forming protein in myrosinase-active seeds. Bellostas and co-workers [22] demonstrated the non-enzymatic and complete degradation of glucosinolates in aqueous solution containing a twofold molar excess of Fe^{2+} ions at pH 5. Our studies show that treating a 1 mM aqueous solution of glucolimnanthin with 10 mM $FeSO_4$ at 90°C for 1 h results in the complete consumption of glucolimnanthin and the formation of nitrile **2** as the only product. The results of these enzyme-mediated and non-enzymatic reactions support the hypothesis that Fe^{2+} ions do not affect the thioglucosidase activity of myrosinase but do interfere with the subsequent Loessen rearrangement, thereby forcing degradation of the intermediate aglycone to the corresponding nitrile and thioamide products [22].

10.3.2 Herbicidal Activity of Glucolimnanthin 1 and Its Individual Degradation Products 2–5

A coleoptile emergence assay was used to determine the herbicidal potency of compounds **1–5**. Seeds of downy brome (*B. tectorum*) were chosen for the assay because downy brome is a prominent weed pest in organic (wheat) farming. Water (zero

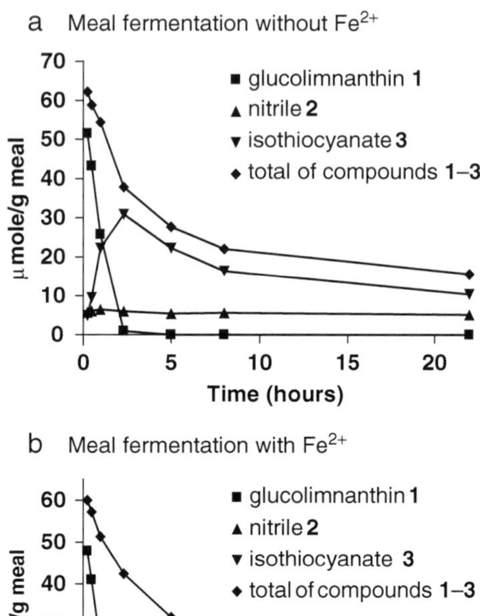

Fig. 10.8 Effect of fermentation on meadowfoam seed meal composition, in the absence (**a**) and presence of 10 mM FeSO$_4$ (**b**). Methods: meadowfoam seed meal (12 g) and untreated meadowfoam seed (cv. Ross) were ground together in a ratio of 100:1 in a coffee grinder for 1 min. Ground mixture (10 g) was mixed with 30 mL of either water or 10 mM FeSO$_4$, vortexed for 1 min, and sonicated for 1 min. At several time points after mixing, 0.1 mL aliquots of the liquid layer were taken and immediately diluted with 0.9 mL methanol, vortexed, centrifuged at 13,000 rpm for 10 min, and analyzed by HPLC as described in Fig. 10.7. Data points represent averages of duplicate incubations

concentration data point) or aqueous solutions of glucolimnanthin at varying concentrations were applied to soil in Petri dishes and then 15 seeds were planted in the soil. In the case of the water-insoluble compounds **2–5**, solutions were prepared in ethanol, mixed with soil, and evaporated in a stream of air. After removal of the solvent, the same volume of water was applied as in the control incubation. Coleoptile emergence was recorded after 7 days of incubation in growth chambers with 8 h day (20°C) and 16 h night periods (15°C) [23]. Dose–response relationships were estimated using logistic regression. After transformation of the logistic regression curves to a linear scale, the relationship between percentage of germinated seeds and the concentration of test compounds showed (partial) S-shaped inhibition curves (Fig. 10.9). Nitrile **2** and thioamide **4** had greater herbicidal potency than glucolimnanthin **1** and isothiocyanate **3**.

10.3.3 Herbicidal Activity of Fermented Meadowfoam Seed Meal

The coleoptile emergence assay was repeated with meadowfoam seed meal and fermented seed meals. The meal materials were mixed with soil at varying doses

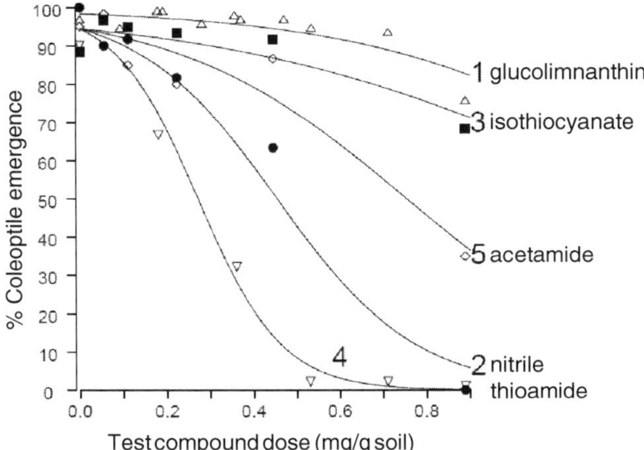

Fig. 10.9 Herbicidal effects of glucolimnanthin and its degradation products determined by a coleoptile emergence assay using downy brome (*Bromus tectorum*) seeds. Method: 15 mL aliquots of aqueous glucolimnanthin solutions were added to 45 g of Walla Walla silt loam soil in 10 cm Petri dishes. For the hydrophobic compounds **2–5**, solutions in ethanol were added to the soil in the Petri dishes and allowed to dry in a stream of air, followed by the addition of 15 mL water. Controls used deionized water or ethanol. Fifteen seeds of *B. tectorum* were planted in each dish. Coleoptile emergence was recorded after 7 days of incubation in growth chambers with 8 h day (20°C) and 16 h night periods (15°C) [23]. Dose–response relationships were estimated using logistic regression with the concentration on a linear scale (adapted from Stevens et al., 2009)

and then downy brome seeds were planted and watered. Coleoptile emergence was recorded after 7 days of incubation. The percentages of downy brome seeds with emerged coleoptiles were subjected to logistic regression. Transformation of the logistic regression curves to a linear scale gave the dose–response relationships presented in Fig. 10.10. The untreated, factory-grade seed meal showed little inhibitory effect on coleoptile emergence at the highest dose (22 mg/g soil), which may explain the variable results obtained with factory-grade meal to control weeds in the field where glucolimnanthin conversion would depend on the presence of soil microorganisms. The fermented seed meals (T1 and T2 seed meals) had greater herbicidal potency than untreated seed meal or the meals treated with either water or 10 mM FeSO$_4$ (T3–T5 seed meals, Fig. 10.10). This finding is in agreement with the enzyme-mediated conversion of glucolimnanthin into the more potent herbicidal agents, isothiocyanate **3**, and especially nitrile **2** (Fig. 10.9). The higher content of nitrile **2** in the seed meal incubated with enzyme-active seeds and FeSO$_4$ translates into greater herbicidal potency (T2 meal, Fig. 10.10) compared to the meal fermented with only enzyme-active seeds (T1 meal).

As pro-bioherbicides, glucosinolates can be utilized to control weeds but the success depends on their conversion into active principles. Agricultural use of untreated seed meal may elicit varying degrees of weed control because environmental factors determine the extent of glucosinolate "bio-activation" in the field [39, 42–45]. Our approach to ferment seed meal using myrosinase (and possibly other enzymes) in

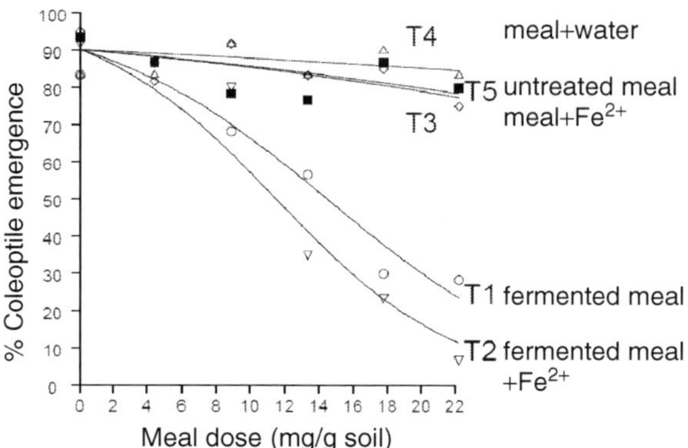

Fig. 10.10 Effect of fermentation of meadowfoam seed meal on herbicidal potency. Methods: seed meal and untreated meadowfoam seeds (*L. alba* ssp. *alba* Benth. cv. Ross) were ground together in a 100:1 ratio in a coffee grinder for 1 min. Ground batches of 10 g each were pooled and mixed with deionized water at a ratio of 3 mL/g of meal and sonicated for 5 min. After incubation for 18 h at room temperature, the mixture was freeze-dried and then reground for 30 s (T1). Preparation of iron-treated meal used the same procedure with 10 mM $FeSO_4$ in place of deionized water (T2). Control incubations included sham-treated meal (ground and freeze-dried but without seeds, T4), iron only (meal plus $FeSO_4$ without seeds, T3), and unaltered meal (ground meal only, not freeze-dried, T5). Coleoptile emergence was measured as in Fig. 10.9 after mixing freeze-dried meal products with soil (adapted from Stevens et al., 2009)

untreated seeds under carefully controlled conditions yields predictable glucosinolate conversion, a product of predictable composition, and greater herbicidal effect than untreated seed meal. Our efforts may ultimately lead to a marketable meadowfoam seed meal product that may increase the profitability of the meadowfoam oil industry.

Acknowledgments Support from the U.S. Department of Agriculture (Grant USDA/CREES 2009-34407-19967), the Oregon Meadowfoam Growers and Natural Plant Products, Inc., Salem (OR), is gratefully acknowledged. The coleoptile emergence assays were conducted by Dr. Steven Machado and Mr. Larry Pritchett (Oregon State University; Columbia Basin Agricultural Research Center, Pendleton, OR). Figures 10.9 and 10.10 were constructed with the help of Dr. Susan Alber (Department of Statistics, Oregon State University). We are also grateful to Dr. Mary Slabaugh (Department of Crop and Soil Science, Oregon State University) for seed samples of *Limnanthes alba* and *L. floccosa*.

References

1. Gentry HS, Miller RW (1965) The search for new industrial crops. IV: prospectus of *Limnanthes*. Econ Bot 19:25–32
2. Jain SK (1986) Domestication of *Limnanthes* (Meadowfoam) as a new oil crop. In Plant domestication induced mutation: Proceedings of an advisory group meeting on the possible use of mutation breeding for rapid domestication of new crop plants, 121–134

3. Miller RW, Daxenbichler ME, Earle FR, Gentry HS (1964) Search for new industrial oils. VIII. The genus *Limnanthes*. J Am Oil Chem Soc 41:167–169
4. Jolliff GD, Tinsley IJ, Calhoun W, Crane JM (1981) Meadowfoam (*Limnanthes alba*): its research and development as a potential new oilseed crop for the Willamette Valley of Oregon. Agricultural Experiment Station, Oregon State University
5. Knapp SJ, Crane JM (1999) Breeding advances and germplasm resources in meadowfoam: a very long chain oilseed. In Janick J (ed) Perspectives on new crops and new uses. ASHS Press, Alexandria, VA, 225–233
6. Crane JM, Knapp SJ (2000) Registration of 'Knowles' meadowfoam. Crop Sci 40:291
7. Knapp SJ, Crane JM, Brunick R (2005) Registration of 'Ross' meadowfoam. Crop Sci 45:407
8. Crane JM, Knapp SJ (2000) Registration of the self-pollinated meadowfoam germplasm line OMF64. Crop Sci 40:1511
9. Crane JM, Knapp SJ (2002) Registration of 'Wheeler' meadowfoam. Crop Sci 42:2208–2209
10. Knapp SJ, Crane JM (1995) Fatty acid diversity of section inflexae *Limnanthes* (meadowfoam). Ind Crops Prod 4:219–227
11. Katengam S, Crane JM, Slabaugh MB, Knapp SJ (2001) Genetic mapping of a macromutation and quantitative trait loci underlying fatty acid composition differences in meadowfoam oil. Crop Sci 41:1927–1930
12. Abbott TP, Wohlman A, Isbell T, Momany FA, Cantrell C, Garlotta DV, Weisleder D (2002) 1,3-Di(3-Methoxybenzyl) thiourea and related lipid antioxidants. Ind Crops Prod 16:43–57
13. Ettlinger MG, Lundeen AJ (1956) The mustard oil of *Limnanthes douglasii* seed, m-methoxybenzylisothiocyanate. J Am Chem Soc 78:1952–1954
14. Stevens JF, Reed RL, Morre JT (2008) Characterization of phytoecdysteroid glycosides in meadowfoam (*Limnanthes alba*) seed meal by positive and negative ion LC-MS/MS. J Agric Food Chem 56:3945–3952
15. Meng Y, Whiting P, Sik V, Rees HH, Dinan L (2001) Limnantheoside C (20-hydroxyecdysone 3-*O*-beta-D-glucopyranosyl-[1→3]-beta-D-xylopyranoside), a phytoecdysteroid from seeds of Limnanthes alba (Limnanthaceae). Zeitschrift fur Naturforschung 56:988–994
16. Sarker SD, Girault JP, Lafont R, Dinan L (1997) Ecdysteroid xylosides from *Limnanthes douglasii*. Phytochemistry 44:513–521
17. Preston-Mafham J, Dinan L (2002) Phytoecdysteroid levels and distribution during development in *Limnanthes alba* Hartw. ex Benth. (Limnanthaceae). Zeitschrift fuer Naturforschung C J Biosci 57:144–152
18. Burow M, Markert J, Gershenzon J, Wittstock U (2006) Comparative biochemical characterization of nitrile-forming proteins from plants and insects that alter myrosinase-catalysed hydrolysis of glucosinolates. FEBS J 273:2432–2446
19. Lambrix V, Reichelt M, Mitchell-Olds T, Kliebenstein DJ, Gershenzon J (2001) The Arabidopsis epithiospecifier protein promotes the hydrolysis of glucosinolates to nitriles and influences Trichoplusia ni herbivory. Plant cell 13:2793–2807
20. Matusheski NV, Swarup R, Juvik JA, Mithen R, Bennett M, Jeffery EH (2006) Epithiospecifier protein from broccoli (*Brassica oleracea* L. ssp. Italica) inhibits formation of the anticancer agent sulforaphane. J Agric Food Chem 54:2069–2076
21. Kissen R, Bones AM (2009) Nitrile-specifier proteins involved in glucosinolate hydrolysis in Arabidopsis thaliana. J Biolog Chem 284:12057–12070
22. Bellostas N, Sorensen AD, Sorensen JC, Sorensen H (2008) Fe2+-catalyzed formation of nitriles and thionamides from intact glucosinolates. J Nat Prod 71:76–80
23. Stevens JF, Reed RL, Alber S, Pritchett L, Machado S (2009) Herbicidal Activity of glucosinolate degradation products in fermented meadowfoam (*Limnanthes alba*) seed meal. J Agric Food Chem
24. Heany RK, Fenwick GR (1993) Methods for glucosinolate analysis. In Waterman PG (ed) Methods in plant biochemistry Academic Press, London

25. Kiddle G, Bennett RN, Botting NP, Davidson NE, Robertson AA, Wallsgrove RM (2001) High-performance liquid chromatographic separation of natural and synthetic desulphoglucosinolates and their chemical validation by UV, NMR and chemical ionisation-MS methods. Phytochem Anal 12:226–242

26. Prestera T, Fahey JW, Holtzclaw WD, Abeygunawardana C, Kachinski JL, Talalay P (1996) Comprehensive chromatographic and spectroscopic methods for the separation and identification of intact glucosinolates. Anal Biochem 239:168–179

27. Lee KC, Chan W, Liang Z, Liu N, Zhao Z, Lee AW, Cai Z (2008) Rapid screening method for intact glucosinolates in Chinese medicinal herbs by using liquid chromatography coupled with electrospray ionization ion trap mass spectrometry in negative ion mode. Rapid Commun Mass Spectrom 22:2825–2834

28. Millan S, Sampedro MC, Gallejones P, Castellon A, Ibargoitia ML, Goicolea MA, Barrio RJ (2009) Identification and quantification of glucosinolates in rapeseed using liquid chromatography-ion trap mass spectrometry. Anal Bioanal Chem 394:1661–1669

29. Zimmermann NS, Gerendas J, Krumbein A (2007) Identification of desulphoglucosinolates in Brassicaceae by LC/MS/MS: comparison of ESI and atmospheric pressure chemical ionisation-MS. Mol Nutr Food Res 51:1537–1546

30. Mohn T, Cutting B, Ernst B, Hamburger M (2007) Extraction and analysis of intact glucosinolates–a validated pressurized liquid extraction/liquid chromatography-mass spectrometry protocol for *Isatis tinctoria*, and qualitative analysis of other cruciferous plants. J Chromatogr 1166:142–151

31. Lee KC, Cheuk MW, Chan W, Lee AW, Zhao ZZ, Jiang ZH, Cai Z (2006) Determination of glucosinolates in traditional Chinese herbs by high-performance liquid chromatography and electrospray ionization mass spectrometry. Anal Bioanal Chem 386:2225–2232

32. Tian Q, Rosselot RA, Schwartz SJ (2005) Quantitative determination of intact glucosinolates in broccoli, broccoli sprouts, brussels sprouts, and cauliflower by high-performance liquid chromatography-electrospray ionization-tandem mass spectrometry. Anal Biochem 343:93–99

33. Matthaus B, Luftmann H (2000) Glucosinolates in members of the family Brassicaceae: separation and identification by LC/ESI-MS-MS. J Agric Food Chem 48:2234–2239

34. Cataldi TRI, Rubino A, Lelario F, Bufo SA (2007) Naturally occurring glucosinolates in plant extracts of rocket salad (*Eruca Sativa* L.) identified by liquid chromatography coupled with negative ion electrospray ionization and quadrupole ion-trap mass spectrometry. Rapid Commun Mass Spectrom 21:2374–2388

35. Skutlarek D, Faerber H, Lippert F, Ulbrich A, Wawrzun A, Buening-Pfaue H (2004) Determination of glucosinolate profiles in Chinese vegetables by precursor ion scan and multiple reaction monitoring scan mode (LC-MS/MS). Eur Food Res Technol 219:643–649

36. Vaughn SF (1999) Glucosinolates as natural herbicides. In Cutler HG, Cutler SJ (eds) Biologically active natural products: agrochemicals. CRC Press, Boca Roton, FL81–91

37. Brown PD, Morra MJ (1995) Glucosinolate-containing plant tissues as bioherbicides. J Agric Food Chem 43:3070–3074

38. Hansson D, Morra MJ, Borek V, Snyder AJ, Johnson-Maynard JL, Thill DC (2008) Ionic thiocyanate (SCN(-)) production, fate, and phytotoxicity in soil amended with Brassicaceae seed meals. J Agric Food Chem 56:3912–3917

39. Gardiner JB, Morra MJ, Eberlein CV, Brown PD, Borek V (1999) Allelochemicals released in soil following incorporation of rapeseed (*Brassica napus*) green manures. J Agric Food Chem 47:3837–3842

40. Vaughn SF, Palmquist DE, Duval SM, Berhow MA (2006) Herbicidal activity of glucosinolate-containing seedmeals. Weed Sci 54:743–748

41. Vaughn SF, Boydston RA, Mallory-Smith CA (1996) Isolation and identification of (3-methoxyphenyl)acetonitrile as a phytotoxin from meadowfoam (*Limnanthes alba*) seedmeal. J Chem Ecol 22:1939–1949

42. Brown PD, Morra MJ (1997) Control of soil-borne plant pests using glucosinolate-containing plants. Adv Agron 61:167–231
43. Borek V, Morra MJ, McCaffrey JP (1996) Myrosinase activity in soil extracts. Soil Sci Soc Am J 60:1792–1797
44. Brown PD, Morra MJ, McCaffrey JP, Auld DL, Williams L (1991) Allelochemicals produced during glucosinolate degradation in soil. J Chem Ecol 17:2021–2034
45. Gimsing AL, Poulsen JL, Pedersen HL, Hansen HC (2007) Formation and degradation kinetics of the biofumigant benzyl isothiocyanate in soil. Environ Sci Technol 41:4271–4276

Chapter 11
Elucidating the Metabolism of Plant Terpene Volatiles: Alternative Tools for Engineering Plant Defenses?

Dorothea Tholl and Sungbeom Lee

11.1 Introduction

Mankind has benefited greatly from the use of plant natural products. To date, the traditional use of phytochemicals as fragrances, flavors, herbal remedies, and food supplements [1, 2] remains as important as the modern development of novel plant-derived pharmaceuticals [3]. Over the past decade, new applications of plant secondary metabolites have emerged based on an expanded knowledge of their physiological and ecological functions. An increasing number of plant metabolites have been investigated for their *in planta* role in the attraction of beneficial organisms or in the defense against pathogens and herbivores [4–10]. These advances have made plant secondary compounds attractive targets for the development of novel, environmentally friendly pest control strategies. Plants have developed different chemical defense strategies that employ various secondary metabolic pathways controlled by specific spatial and temporal regulatory regimes. Direct defense mechanisms can be based on premade toxic or deterrent volatile and nonvolatile metabolites that are stored in specialized cells and released or activated upon attack [11–13]. Alternatively, compounds with direct defense activities can be synthesized *de novo* in a response induced systemically or locally at the site of herbivore damage or pathogen invasion [14]. One of the most sophisticated defense mechanisms of plants is the ability to "call for help" from other organisms. Such indirect defense responses rely primarily on the release of herbivore-induced plant volatiles (HIPVs), which attract parasitoids and predators of herbivores [15–18] (Fig. 11.1). This tritrophic interaction has been demonstrated to occur in a variety of plant species including important crops such as *Brassica* [19], maize [20], rice [21–23], tomato [24], and cucumber [25]. The volatile blends of compounds emitted upon herbivore damage consist in large part of terpenes mixed with the benzenoid volatile, methyl salicylate, and products of the lipoxygenase

D. Tholl (✉)
Department of Biological Sciences, Virginia Tech, Blacksburg, VA 24061, USA
e-mail: tholl@vt.edu

D.R. Gang (ed.), *The Biological Activity of Phytochemicals*, Recent Advances in Phytochemistry 41, DOI 10.1007/978-1-4419-7299-6_11,
© Springer Science+Business Media, LLC 2011

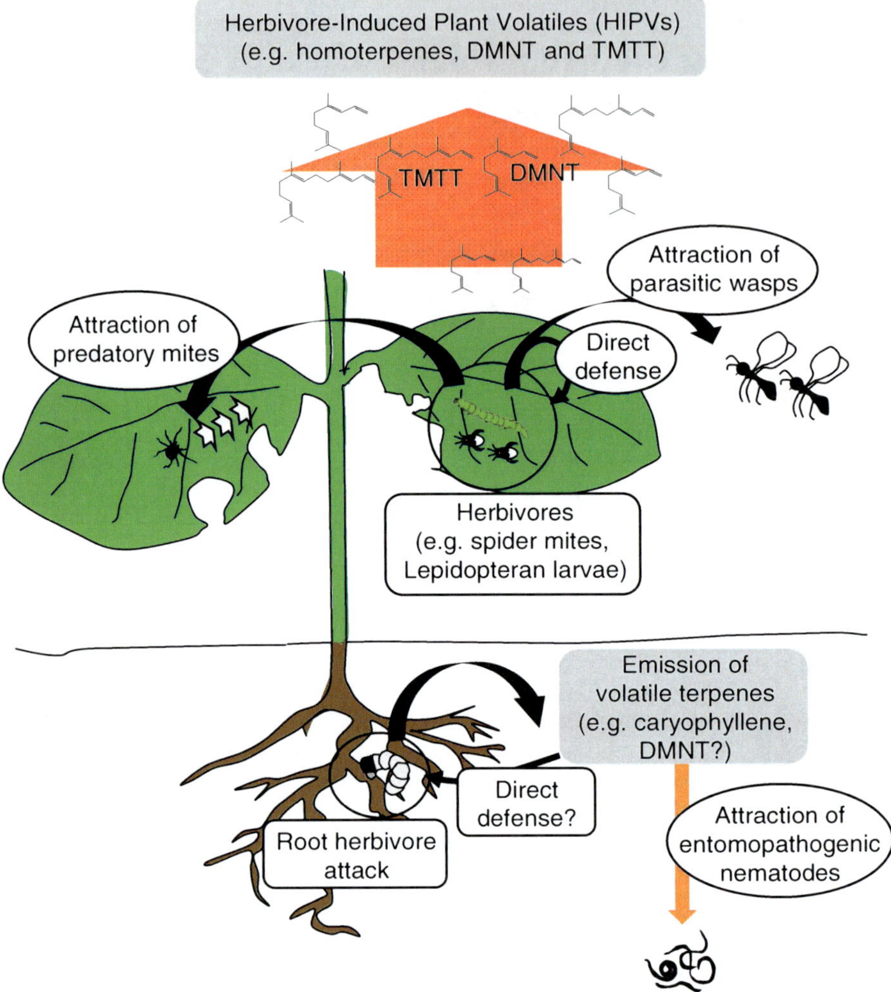

Fig. 11.1 Simplified scheme of the emission and function of herbivore-induced plant volatiles in plant defense. DMNT, 4,8-dimethyl-1,3,7-nonatriene); TMTT, 4,8,12-trimethyltrideca-1,3,7, 11-tetraene

pathway such as green leaf volatiles and methyl jasmonate. The formation and function of these volatiles and their potential use in metabolic engineering of plant defenses have received much attention by biochemists and ecologists alike. Therefore, recent research interests have been focused on elucidating the biosynthesis and regulation of HIPVs with the goal of deciphering their individual or synergistic functions in indirect defense. Here, we describe current advances in the analysis of HIPV biosynthesis with a focus on homoterpene formation in *Arabidopsis thaliana*. We address approaches and future challenges in metabolic engineering of homoterpenes and other volatile terpenes for agricultural pest control.

11.2 Targeting Volatile Terpene Biosynthesis for Manipulating Indirect Plant Defenses

Growing interest in environmentally sound crop protection has stimulated research on enhancing chemical defenses in crop plants. In crucifers, for example, efforts have been made to change the levels and types of glucosinolates, the predominant secondary defense metabolites in the Brassicaceae family, by overexpressing key enzymes of the glucosinolate biosynthetic pathways and transcription factors controlling these pathways [12, 26, 27]. Volatiles that function in direct and indirect plant defense have gained increasing recognition for their potential use in biological pest control. In particular, enhancing the emission of volatiles by genetic engineering is thought to improve the effectiveness of attracting natural enemies of herbivores. The fact that HIPVs can increase plant indirect defenses has been demonstrated impressively by direct applications of volatile monoterpenes and sesquiterpenes in natural ecosystems [16]. Also, integrated pest management strategies, the so-called push–pull applications, have been employed successfully to lure herbivore enemies to crops by intercropping with plants that emit high levels of attractive volatiles [28, 29].

Recent successes in the genetic manipulation of plant volatile defenses have been reported [30]. Special progress toward engineering indirect defenses in crop plants has been made by studies on the biosynthesis of volatile terpenes in maize and rice. Maize leaves emit a complex mixture of insect-induced volatiles consisting mostly of monoterpenes and sesquiterpenes [31]. The volatile blend attracts parasitic wasps such as *Cotesia marginiventris*, which oviposit into the insect larvae. The parasitized larvae have reduced feeding activity, which increases plant fitness [32]. Transgenic *Arabidopsis* plants expressing a maize sesquiterpene synthase responsible for the formation of the predominant herbivore-induced sesquiterpene volatiles were shown to attract *C. marginiventris* females that had been trained to associate these compounds with their larval host [10]. In rice, similar results were achieved by expressing a terpene synthase, which produces the sesquiterpene (*E*)-β-caryophyllene, a major terpene volatile emitted from rice [33]. Transgenic rice plants with increased emission rates of (*E*)-β-caryophyllene were more attractive to *Anagrus nilaparvatae*, an egg parasitoid of the rice brown planthopper *Nilaparvata lugens*, than non-transgenic control plants [33].

A major advance to enhance indirect chemical defenses in the roots of crop plants has been made by manipulating emissions of volatile terpenes from maize roots. This approach is based on the discovery that feeding by the western corn rootworm, *Diabrotica virgifera virgifera*, causes roots of certain maize cultivars to release (*E*)-β-caryophyllene. This sesquiterpene attracts the generalist entomopathogenic nematode, *Heterorhabditis megidis*, that can parasitize *Diabrotica* larvae [5] (Fig. 11.1). Interestingly, North American maize cultivars lack the ability to produce (E)-β-caryophyllene and are less attractive to nematodes [34] than European cultivars. By expressing an (E)-β-caryophyllene synthase from oregano in the American cultivars, the resulting transgenic plants gained the ability to attract nematodes and showed significantly reduced root damage when tested under field conditions [35].

In line with these successful results, increasing emphasis has been placed on characterizing the function of homoterpene volatiles in plant defense. Homoterpenes such as the C_{11}-homoterpene DMNT (4,8-dimethylnona-1,3, 7-triene) and its higher-molecular-weight homolog, the C_{16}-homoterpene TMTT (4,8,12-trimethyltrideca-1,3,7,11-tetraene), are irregular terpenes, which are among the most common volatile compounds emitted from plant foliage upon herbivore attack [36–39]. Several studies have provided strong support for the role of DMNT and TMTT in the attraction of herbivore predators. Predatory mites such as *Phytoseiulus persimilis* are attracted to DMNT when offered as an individual compound [40, 41] (Fig. 11.1). While TMTT is not attractive to predatory mites as a pure compound, De Boer et al. [42] demonstrated that TMTT influenced the foraging behavior of these predators when emitted from spider mite-infested lima bean leaves in the presence of other induced volatiles. A similar response by predatory mites was observed on tomato plants, which released larger amounts of TMTT upon spider mite attack than control plants [24]. Also, treatments of lima bean with the terpenoid pathway inhibitor fosmidomycin, which severely reduced the emission of homoterpenes, led to reduced attraction of predatory mites [43]. Finally, the attractive function of homoterpene volatiles was corroborated by recent studies with genetically engineered *Arabidopsis* plants [6]. Transgenic plants, which emitted DMNT together with its precursor $(3S)$-(E)-nerolidol constitutively or in response to treatment with jasmonic acid, were more attractive to predatory mites in olfactometer experiments and under semi-natural conditions than non-emitting wild-type controls. The study also showed that $(3S)$-(E)-nerolidol by itself can attract predatory mites similar to the effect of the analogous C_{10}-alcohol linalool produced in transgenic potato [44].

The function of homoterpenes in plant defense goes beyond just the attraction of insect predators. For example, emission of TMTT has been discussed with respect to the repulsion of aphids [45]. Moreover, homoterpene volatiles were found to elicit defense responses in plants as demonstrated by the induced expression of defense genes in lima bean when exposed to TMTT [46]. Therefore, homoterpenes might also play a role in intra-plant defense signaling and in plant–plant interactions by priming non-infested neighboring plants for defense responses. Priming crop plants with transgenic homoterpene emitters in the field could add another strategy to improve plant resistance.

A more detailed characterization of homoterpenes as volatile signals in plant defense and their potential application in metabolic engineering of indirect defenses has been restricted by a limited understanding of homoterpene biosynthesis and its regulation. The genetic resources of *Arabidopsis*, which emits homoterpenes above and below ground, have become valuable tools for achieving this goal.

11.3 Arabidopsis: A Small Weed and Its Genetic Resources for Volatile Terpene Biosynthesis

Despite its short life cycle, *Arabidopsis thaliana* maintains a rich array of secondary metabolism that is evident in the formation of glucosinolates, flavonoids, and terpenes. Substantial progress has been achieved in elucidating the biosynthesis of volatile terpene metabolites using *Arabidopsis* as a model system [47–50]. Terpene volatiles are synthesized in all major organs of the *Arabidopsis* plant by enzymes called terpene synthases (TPSs). Generally, TPSs catalyze the formation of C_{10}-monoterpenes, C_{15}-sesquiterpenes, and C_{20}-diterpenes in a single enzymatic step from the central terpene precursors, geranyl diphosphate (GPP), farnesyl diphosphate (FPP), and geranylgeranyl diphosphate (GGPP) [51] (Fig. 11.2). The prenyldiphosphate substrates are derived from the C_5-precursor IPP (isopentenyl

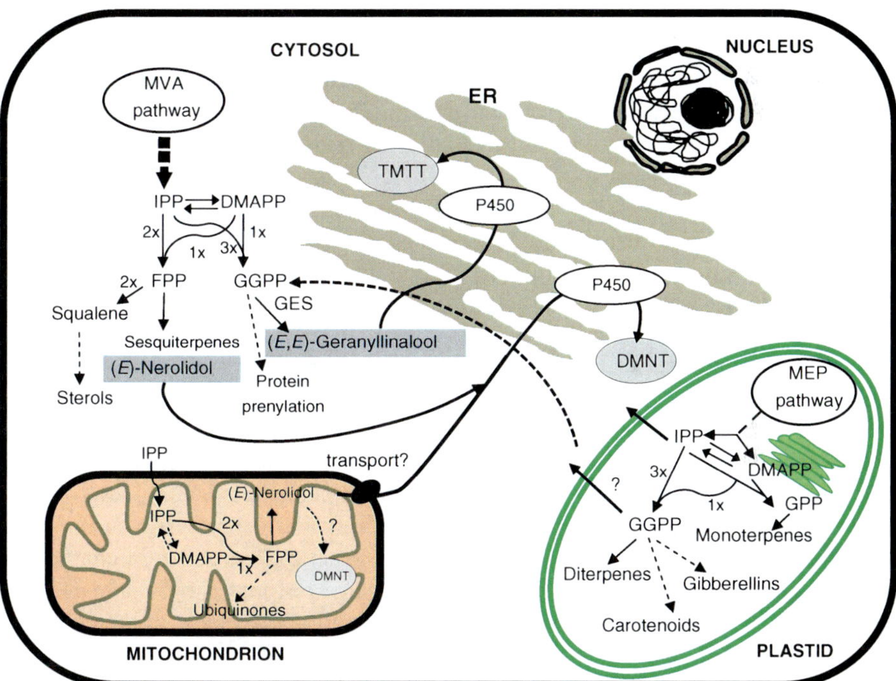

Fig. 11.2 Terpene biosynthesis in the plant cell. A model is shown for the biosynthetic pathways and their subcellular compartmentalization in the formation of the volatile homoterpenes DMNT (4,8-dimethyl-1,3,7-nonatriene) and TMTT (4,8,12-trimethyltrideca-1,3,7,11-tetraene). MVA, mevalonate; MEP, methylerythritol phosphate; ER, endoplasmic reticulum; P450, cytochrome P450 monooxygenase; GES, geranyllinalool synthase; IPP, isopentenyl diphosphate; DMAPP, dimethylallyl diphosphate; GPP, geranyl diphosphate; FPP, farnesyl diphosphate and GGPP, geranylgeranyl diphosphate. The putative enzymatic conversion of (*E*)-nerolidol to DMNT in the mitochondria and a possible export of GGPP from plastids to the cytosol are indicated by question marks

diphosphate) and its isomer DMAPP (dimethylallyl diphosphate), which are formed in the cytosol by the mevalonate (MVA) pathway and in plastids via the methylerythritol phosphate (MEP) pathway (Fig. 11.2). While monoterpenes and diterpenes are predominantly produced in plastids, sesquiterpenes are synthesized mainly in the cytosol [51]. As a result of gene duplication and sequence divergence, the *Arabidopsis* genome contains a large family of 32 terpene synthase (*AtTPS*) genes [52, 53]. Many of these genes show distinct transcription patterns restricted to particular organs and cell types of flowers (7 *TPS* genes) and roots (14 *TPS* genes), which are most likely the result of functional specialization under selection pressures in interaction with other organisms. For example, the cineole monoterpene synthase gene *AtCin* is specifically expressed in root epidermal cells where it may function in the defense against invading microbial pathogens [54, 55]. Other root-specific terpene synthases show transcript profiles confined to the vascular tissue or the cortex and endodermis according to high-resolution gene transcript maps in *Arabidopsis* roots [56, 57].

In addition to the regular C_{10}-, C_{15}-, and C_{20}-terpenes, *Arabidopsis* produces both homoterpenes DMNT and TMTT. In most plants, these compounds are emitted together from the foliage upon herbivore attack. However, in *Arabidopsis*, DMNT is detected exclusively in roots following treatment with jasmonic acid (Table 11.1) or elicitation by the root-rot pathogen *Pythium irregulare* (Huh et al., unpublished results), indicating an unexpected role of DMNT in root defense. TMTT is synthesized *de novo* in *Arabidopsis* leaves in response to fungal elicitor treatment or feeding damage by the crucifer specialists *Plutella xylostella* and *Pieris rapae* [58, 59] (Table 11.1), and it is emitted as the predominant volatile compound together with the monoterpene β-ocimene, the sesquiterpene (*E,E*)-α-farnesene, and methyl salicylate (in the Columbia ecotype). The emitted volatiles were shown to attract the herbivore-parasitizing wasp *Cotesia rubecula*, supporting their role in indirect defense [59]. Moreover, the parasitization of *P. rapae* larvae by *C. rubecula* resulted in an increase in plant fitness as measured by seed production [60]. The specific function of TMTT in the attraction of *C. rubecula* and other parasitoids has not yet been fully resolved and will require bioassays with plants lacking or overproducing TMTT based on the knockout or overexpression of the genes involved in TMTT biosynthesis. Functional genomics and reverse genetic approaches in characterizing terpene biosynthesis genes in *Arabidopsis* have become particularly useful in elucidating the biosynthetic machinery in the formation of volatile homoterpenes.

11.4 How Does Arabidopsis Assemble Homoterpene Volatiles?

11.4.1 Formation of Alcohol Precursors

Elucidating the enzymatic steps in the formation of volatile homoterpenes has been a subject of continual investigation. C_{11}- and C_{16}- homoterpenes are acyclic terpenes that are produced by degradation of C_{15}- and C_{20}- precursors. Initial studies by Boland and co-workers [61] with deuterium-labeled precursors administered to

Table 11.1 Enzymes and genes with proposed or proven function in herbivore-induced biosynthesis of DMNT and TMTT in model and crop plants[a]

Species	Homoterpene emission		Enzyme	Gene	References (referring to enzyme or gene)
	DMNT	TMTT			
Arabidopsis thaliana	Roots, JA	Leaves, *Pieris rapae* [59]; *Plutella xylostella* [58]	(*E*,*E*)-geranyllinalool synthase	*GES/TPS04* (At1g61120)	[58]
			Root-expressed (*E*)-nerolidol synthases	–	Unpublished results
			Root- and leaf-expressed P450s	–	Unpublished results
Zea mays	Leaves, *Spodoptera littoralis* [31]	Leaves, *Spodoptera littoralis* [31]	(3*S*)-(*E*)-nerolidol synthase	–	[62]
			(3*R*)-(*E*)-nerolidol/(*E*)-β-farnesene/(*E*,*E*)-farnesol synthase	*tps1*	[64]
Medicago truncatula	Leaves, *Spodoptera exigua* [76]	Leaves, *Spodoptera exigua* [76]	(3*S*)-(*E*)-nerolidol/ geranyllinalool synthase	*MtTPS3*	[76]
Phaseolus lunatus	Leaves, *Tetranychus urticae* [38, 63], *Spodoptera littoralis* [96]	Leaves, *Tetranychus urticae* [38, 63], *Spodoptera littoralis* [96]	(3*S*)-(*E*)-nerolidol synthase	–	[63]
Cucumis sativus	Leaves, *Tetranychus urticae* [63]	Leaves, *Tetranychus urticae* [63]	(3*S*)-(*E*)-nerolidol synthase	–	[63]
Solanum lycopersicum	Leaves, *Tetranychus urticae* [24]	Leaves, *Tetranychus urticae* [24] (predominant volatile)	Geranyllinalool synthase	–	[37]

[a]Nerolidol synthases of non-homoterpene emitting plants are not shown. JA, jasmonic acid.

excised lima bean plantlets treated with jasmonate indicated that DMNT and TMTT are derived from the tertiary sesquiterpene and diterpene alcohols (*E*)-nerolidol and (*E,E*)-geranyllinalool, respectively. Based on these experiments, it became evident that the enzymatic conversion of the prenyldiphosphates FPP and GGPP into (*E*)-nerolidol and (*E,E*)-geranyllinalool, respectively, is the first dedicated step in homoterpene formation, which is supposed to proceed via hydrolysis of the diphosphate moiety and allylic rearrangement to give a tertiary alcohol with a terminal olefin (Fig. 11.3). In maize, feeding damage by the lepidopteran, *Spodoptera littoralis*, induces a nerolidol synthase activity, by which FPP is converted into the (3*S*)-(*E*)-nerolidol enantiomer [62] (Table 11.1). Similar (3*S*)-(*E*)-nerolidol synthase activities in correlation with DMNT emissions were demonstrated in leaves of lima bean and cucumber after feeding damage by the spider mite *Tetranychus urticae* [63] (Table 11.1). In both cases, (*E*)-nerolidol was not detected or found only in traces in the volatile blend, indicating a rapid conversion of this terpene alcohol into DMNT. Despite finding nerolidol synthase activities, the identification of the respective genes involved in DMNT formation has lagged behind. Schnee et al. [64] isolated a terpene synthase (*tps1*) from maize, whose corresponding enzyme produces (3*R*)-(*E*)-nerolidol as a major product and is inducible by insect herbivory (Table 11.1). TPS1 most likely functions together with other (*E*)-nerolidol synthases

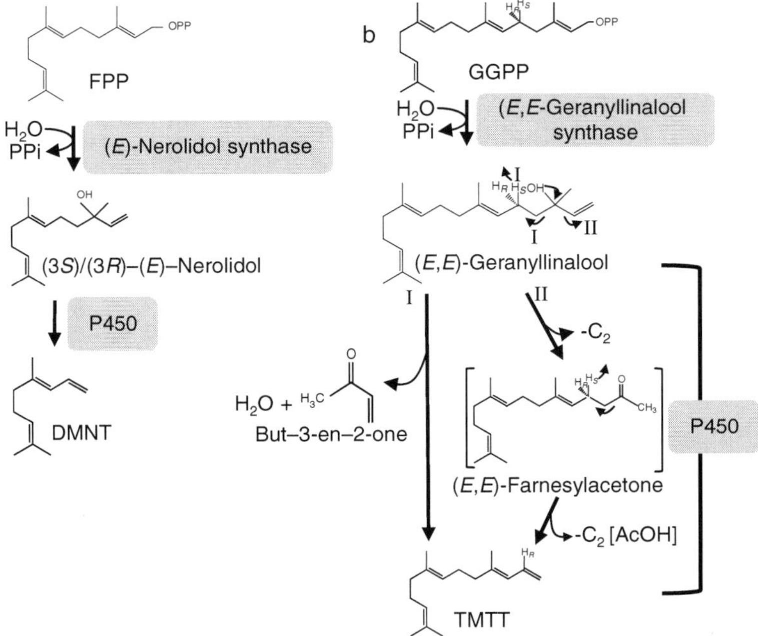

Fig. 11.3 Biosynthesis of the homoterpenes (**a**) DMNT (4,8-dimethyl-1,3,7-nonatriene) and (**b**) TMTT (4,8,12-trimethyltrideca-1,3,7,11-tetraene). (**b**) Two putative reaction mechanisms for the conversion of (*E,E*)-geranyllinalool to TMTT catalyzed by a P450 enzyme activity are shown. FPP, farnesyl diphosphate; GGPP, geranylgeranyl diphosphate

since both the (3*R*) and the (3*S*) enantiomer of (*E*)-nerolidol can be converted into DMNT when administered to maize leaves [62, 64]. A reverse genetics analysis of root-expressed *TPS* genes in *Arabidopsis* has revealed the existence of two genes, whose recombinant enzymes produce (*E*)-nerolidol as a side product and may be in part involved in the formation of DMNT possibly together with other root-specific metabolic routs (Sohrabi, Huh, et al., unpublished results).

Consistent with the demonstration of nerolidol as a precursor in DMNT synthesis, Ament et al. [37] reported a stringent correlation between the biosynthesis of the diterpene alcohol (*E,E*)-geranyllinalool (Table 11.1) and the formation of TMTT in tomato leaves in response to treatment with jasmonic acid. While a gene encoding a geranyllinalool synthase has not yet been isolated from tomato, in *Arabidopsis* a single gene (*TPS04*, *GES*) has been identified that is responsible for the *de novo* formation of geranyllinalool as the central precursor in herbivore-induced TMTT biosynthesis [58] (Table 11.1). Transcription of *GES* and the biosynthesis of geranyllinalool and TMTT are induced locally by feeding of the specialist *P. xylostella* and by treatment with the fungal elicitor alamethicin or the jasmonate mimic coronalon. Regulation of *GES* transcription is dependent on the octadecanoid-dependent signaling pathway and is not modified by salicylic acid or ethylene [58].

The *Arabidopsis* GES protein shares 40% sequence identity with two characterized linalool synthases from *Clarkia breweri* and *Clarkia concinna* [65], which together belong to the TPS-f family (Fig. 11.4). Linalool synthases catalyze a reaction analogous to that of GES by converting the 10-carbon substrate GPP into

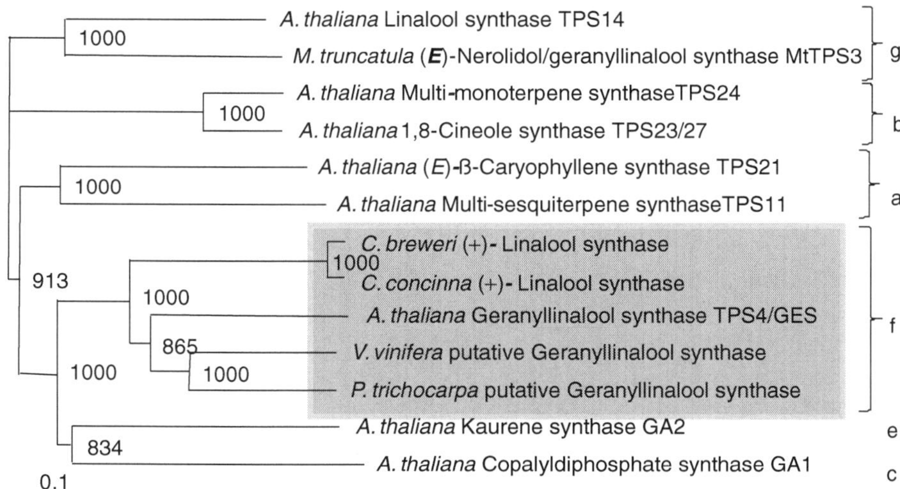

Fig. 11.4 Sequence comparison of *Arabidopsis* geranyllinalool synthase with other *Arabidopsis* TPS and putative or characterized linalool and geranyllinalool synthases from *Clarkia, Medicago*, poplar and grape. A neighbor-joining tree generated from an amino acid sequence alignment (ClustalX) of 13 TPS proteins is shown. Numbers are bootstrap values higher than 800 out of 1,000 replicates. TPS proteins belong to different families as indicated by *letters*. *V. vinifera* putative GES, Acc. 002268557; *P. trichocarpa* putative GES, Acc. 002305640

linalool. Strikingly, *GES* and the *Clarkia* linalool synthase genes have almost identical gene structures, which raises the question as to whether linalool synthases of this particular lineage may have evolved from geranyllinalool synthases or vice versa by mutations changing the size of the binding pocket for the substrates GPP or GGPP at the catalytically active site of the enzymes.

11.4.2 Oxidative Degradation of Geranyllinalool to TMTT

In the second step of the homoterpene biosynthetic pathway, (*E*)-nerolidol and (*E*,*E*)-geranyllinalool are converted to their corresponding homoterpene end products by oxidative degradation (Fig. 11.3). Studies with stable isotope precursors administered to lima bean leaves demonstrated that the transformation of the tertiary alcohols into homoterpenes proceeds via an oxidative C–C bond cleavage reaction with an exclusive syn-elimination of the oxygen-carrying group together with an allylic hydrogen atom at the β-carbon atom [66] (Fig. 11.3). This type of reaction is catalyzed by cytochrome P450 monooxygenases (P450s) in other metabolic pathways such as the dealkylation of steroids [67] and the formation of the furanocoumarin psoralen by oxidative degradation of marmesin [68]. The role of P450s in the formation of TMTT in *Arabidopsis* has recently been suggested by Bruce et al. [45]. Based on these findings, our efforts have focused on searching for P450 genes involved in DMNT and TMTT biosyntheses in *Arabidopsis*. Studies on triterpenoid biosynthesis in *Arabidopsis* have provided important examples of metabolic pathways assembled from P450 genes that are tightly coexpressed with other biosynthetic enzymes of the same pathway [69]. Under the assumption that TMTT is formed by an induced coordinated expression of GES with a P450 enzyme, we have screened the recently established *Arabidopsis* BAR (http://bar.utoronto.ca/) and ATTED (http://atted.jp/) co-expression databases using *GES* as bait. Among several selected P450 gene candidates, a single gene has been identified that is responsible for the conversion of geranyllinalool to TMTT (Lee et al., unpublished results). Our findings most likely exclude other enzymes to be involved in the oxidative degradation of geranyllinalool such as peroxidases that were suggested by Mercke et al. [25] to catalyze the conversion of nerolidol to DMNT in cucumber.

The reaction mechanism for the formation of TMTT from geranyllinalool remains to be investigated in greater detail. One possible reaction could proceed via the cleavage of geranyllinalool into TMTT and the C_4-ketone but-3-en-2-one (Fig. 11.3b, pathway I) although the latter compound has not been detected so far. Alternatively, geranyllinalool could undergo a sequential cleavage of two C_2 fragments with the formation of the C_{18}-ketone farnesylacetone as an intermediate (Fig. 11.3b, pathway II). The production of a C_{13}-geranylacetone intermediate in DMNT biosynthesis was suggested by Boland et al. [61] since small amounts of geranylacetone were found in floral and leaf volatile blends of DMNT-emitting plants and labeled geranylacetone was converted into DMNT when administered to lima bean leaves.

11.4.3 Subcellular Compartmentalization of Homoterpene Biosynthesis

To successfully approach the metabolic engineering of homoterpenes, the subcellular localization of the respective biosynthetic enzymes will require attention. The formation of diterpenes is believed to occur predominantly in plastids based on the characterization of diterpene synthases containing plastidial transit peptide sequences [51, 70]. In contrast, *Arabidopsis* GES is missing a plastidial transit peptide because of the absence of three N-terminal exons, which are present in the plastid-localized, most closely related diterpene synthases, copalyl diphosphate synthase and kaurene synthase, in the *Arabidopsis* TPS family [58] (Fig. 11.4). Transgenic *GES* plants constitutively expressing a functionally active GES-yellow fluorescent (YFP) fusion protein further confirmed that GES is not targeted to the plastid but resides in the cytosol or the endoplasmic reticulum (ER) (Fig. 11.2) [58]. It is likely that a pool of GGPP, the substrate for conversion into geranyllinalool, is present in the cytosol or ER since two *Arabidopsis* GGPP synthases have a localization pattern similar to that observed for GES [71]. GGPP may also be transported from plastids to the cytosol as was recently suggested as a means to supply substrate for the prenylation of proteins in tobacco BY-2 cells [72]. Our current model predicts that the formation of geranyllinalool and TMTT both occur at the ER or that geranyllinalool is produced in the cytosol and then converted by a P450 enzyme localized at the ER (Fig. 11.2). Additional studies of both enzymes should give more detailed information about their subcellular localization and possible interaction.

Similar to the formation of TMTT, the biosynthesis of (*E*)-nerolidol and DMNT is predicted to occur in the cytosol and at the ER (Fig. 11.2). (*E*)-Nerolidol synthases from snapdragon (AmNES/LIS-1) and strawberry (FaNES1) were localized in the cytosol [73, 74]. However, DMNT formation may involve more than one subcellular compartment. Indications that nerolidol could be produced in the mitochondria came from subcellular localization studies of a FaNES1-related nerolidol synthase protein [73]. Furthermore, the constitutive expression of strawberry FaNES1 targeted to the mitochondria in *Arabidopsis* led to the emission of DMNT in leaves of transgenic plants [6]. No DMNT formation was observed when the gene was expressed without the mitochondrial transit peptide. The studies suggest that (*E*)-nerolidol, when produced from FPP in the mitochondria, could be either converted to DMNT in the same compartment or transported to the cytosol and the ER and metabolized into DMNT (Fig. 11.2). The fact that terpene biosynthetic pathways can occur in several cell organelles has been demonstrated in peppermint, where limonene is synthesized in plastids but is hydroxylated by a P450 at the ER [75]. Determination of the subcellular location of the (*E*)-nerolidol synthases in *Arabidopsis* roots will provide further insight into the compartmentalization of the DMNT biosynthetic pathway.

11.5 Metabolic Engineering of Homoterpene Volatiles

Current progress in isolating the genes responsible for the biosynthesis of homoterpenes in *Arabidopsis* will provide tools for engineering homoterpene volatiles in other plants. Identification of functionally equivalent genes from other species has so far been limited. Two (*E*)-nerolidol synthases, FaNES1 and AmNES/LIS-1, from strawberry and snapdragon, respectively, were isolated, whose primary function is the formation of cytosolic (*E*)-nerolidol in fruits and flowers [73, 74]. Both proteins belong to the TPS-g family and also convert GPP to linalool. The genomes of poplar and grape contain at least one gene locus with approximately 50% amino acid sequence identity to *Arabidopsis* GES, indicating that they may encode geranyllinalool synthases (Fig. 11.4). Recently, a geranyllinalool synthase (MtTPS3) was characterized from *Medicago truncatula* [76] (Table 11.1). This enzyme can also produce (*E*)-nerolidol from FPP and may be responsible for synthesizing both (*E*)-nerolidol and geranyllinalool precursors in DMNT and TMTT formation. MtTPS3 shares only 20% sequence identity with *Arabidopsis* GES and is more closely related to the *Arabidopsis* linalool synthase TPS14 of the TPS-g family (Fig. 11.4), suggesting that this enzyme, as with many other terpene synthases, has arisen independently as a product of convergent evolution. A similar scenario can be envisioned for P450s catalyzing the oxidative degradation of nerolidol or geranyllinalool, which makes screening and identification of these genes in plants other than *Arabidopsis* more challenging, especially if large populations of gene knockout plant lines are not readily available.

Enhancing the formation of homoterpene volatiles will require a better understanding of the regulation of each step of the homoterpene biosynthetic pathway. Studies in maize and cucumber indicated that the rate-limiting step in the herbivory-induced release of DMNT is the formation of (*E*)-nerolidol [62, 63]. In *Arabidopsis*, TMTT biosynthesis is primarily regulated by transcription of *GES*. Transgenic *Arabidopsis* plants with constitutive expression of *GES* emit TMTT at rates that can be as high as those observed from elicitor-induced wild-type plants [58]. The formation of TMTT in these transgenic lines is possible because of a basal expression of the P450 enzyme catalyzing the oxidative degradation of geranyllinalool. Transgenic plants that constitutively express this P450 enzyme activity produce larger amounts of TMTT only when treated with an elicitor that induces *GES* expression (Lee et al., unpublished results). Despite the regulatory role of GES, its constitutive expression as a single gene can lead to reduced plant performance, eventual instability of transgenic lines, and the temporary formation of lesions at the cotyledon stage most likely because of toxic or stress effects induced by geranyllinalool that accumulates in these plants [58]. Bleaching effects due to the depletion of isoprenoid precursors for carotenoid biosynthesis were not observed as was reported for the expression of strawberry FaNES1 in potato plastids [44]. Further analysis showed that under conditions of high GES activity, the oxidative degradation step can become rate-limiting, even when this step is induced by elicitor treatment [58]. Therefore, the expression of both, *GES* and P450 genes, under the control of a strong promoter will be required to convert geranyllinalool efficiently to TMTT and minimize growth inhibitory effects. This approach may also be applicable for engineering the formation of DMNT. *Arabidopsis* plants expressing strawberry FaNES1 targeted to the mitochondria were shown to emit (*E*)-nerolidol. Depending on the transgenic line, conversion to DMNT was observed with or without treatment with jasmonic acid, and this reaction is most likely catalyzed by a P450 enzyme identical or similar to that involved in TMTT biosynthesis. An efficient metabolite flow toward DMNT could be achieved by the strong constitutive expression of this P450 activity. This approach can also help avoid a possible redirection of the alcohol intermediate to glycosylated end products. Glycosylation can critically reduce the formation of free terpene alcohols as was shown previously in attempts to engineer the formation of linalool in petunia, *Arabidopsis*, and potato [44, 50, 77].

The phenotypic effects observed under the constitutive expression of *GES* may also be avoided by the use of inducible promoters. Transgenic *Arabidopsis* expressing *GES* controlled by the alcohol-inducible *AlcA* promoter did not show the lesion phenotype observed under constitutive expression of *GES* [58]. Recent experiments with wound- and insect-inducible promoters such as that of the potato proteinase inhibitor II have been reported [78]. The use of strong inducible promoters that respond to herbivore-feeding may not only minimize toxic effects or costs associated with the constitutive formation of a volatile compound but can also provide the advantage that parasitoids associate volatiles only with the presence of feeding damage by the respective host.

Targeting insect-induced transcription factors controlling homoterpene and other volatile terpene metabolic pathways is undoubtedly an important goal of future efforts to manipulate volatile biosynthesis in plants. Steady progress has been made in identifying transcription factors involved in regulating single or multiple secondary metabolic pathway genes such as in the formation of phenylalanine-derived anthocyanins and lignins, indole alkaloids, glucosinolates, and benzenoid volatiles of floral scent [26, 79–83]. By contrast, only two transcription factors have been reported to date, which are involved in regulating single or multiple enzymatic steps in terpene secondary metabolism. GaWRKY1 regulates the expression of a δ-cadinene terpene synthase in cotton (*CAD1-A*) in response to fungal elicitor treatment [84] and the bZIP transcription factor OsTGAP1 functions as a key regulator of the coordinated transcription of genes involved in induced diterpene production in rice [85]. Exploiting the comprehensive *Arabidopsis* transcriptome databases has allowed the identification of transcription factors that are co-regulated with genes of specific secondary metabolic pathways [86]. Similar coexpression analyses using the identified homoterpene biosynthesis genes will help pinpoint transcriptional activators or repressors in homoterpene volatile formation.

Producing a target metabolite via genetic engineering typically requires the availability of sufficiently large precursor metabolite pools. Substrate availability in different subcellular compartments and metabolic cross-talk between these compartments differs depending on the plant species. Several previous attempts to engineer the formation of volatile terpenes have been hampered by small and highly regulated metabolite pools shared with primary metabolic pathways such as that for sterol biosynthesis [50, 87, 88]. Overexpression of the regulatory enzymes of the MVA or MEP pathways, HMG-CoA reductase (HMGR), and deoxyxylulose-phosphate synthase (DXS), together with FPP or GGPP synthases, represents a possible strategy to increase carbon flux toward the production of terpenes such as DMNT or TMTT. A similar approach expressing HMGR with FPP synthase and a patchoulol synthase has been applied successfully to engineer sesquiterpene volatiles from cytosolic FPP in tobacco [89]. However, terpene yields in these studies were increased dramatically when the precursor metabolite flow was redirected by targeting FPPs and patchoulol synthase to plastids as their nonnative subcellular compartment, to circumvent cytosolic regulatory constraints [89]. Similarly, strawberry FaNES1, when targeted to the mitochondria, produced (*E*)-nerolidol from the FPP in this compartment for subsequent conversion into DMNT [6]. Using an endogenous mitochondrial GGPP substrate pool or increasing such a pool may also be a suitable way to enhance emissions of TMTT if cytosolic production of geranyllinalool is tightly regulated.

Expressing nerolidol or geranyllinalool synthases for the production of homoterpenes may lead to the formation of other terpenes such as linalool. Such a scenario was described for the engineering of volatile terpenes in tomato fruits where expression of an α-zingiberene sesquiterpene synthase caused the formation of several monoterpenes [90]. This outcome is possible when the heterologously expressed enzyme accepts more than a single prenyldiphosphate substrate and encounters these substrates in the subcellular compartment where it resides. Selecting enzymes with high substrate specificity can improve the specificity of the volatile output.

11.6 Conclusions and Future Challenges

Much progress has been made in elucidating the biosynthesis of plant volatile terpenes. Molecular tools have been used successfully to improve terpene aroma and scent in fruits or flowers [91, 92] and are now employed for engineering volatiles to improve biological pest control. Future challenges will arise in testing the efficacy of engineered volatile emissions beyond lab settings in greenhouse and field experiments. Previous studies on tritrophic interactions demonstrated that the role of volatile terpenes, especially that of TMTT in the attraction of predatory mites, depends on the presence and composition of other volatiles [42]. Therefore, the function of individual volatiles such as homoterpenes together with other compounds of an emitted odor mixture needs to be further corroborated. This could be achieved, for example, by using a series of transgenic plant lines, in which homoterpene emissions are up- or downregulated relative to the other constituents of the volatile blend. Adjusting volatile blends to the preferences of a particular parasitoid or predator will be critical for optimal attraction of these biocontrol agents. Moreover, how the efficiency of engineered volatiles is influenced by abiotic stress factors or upon multispecies attack [93] needs to be investigated in greater detail.

Other aspects of implementing indirect volatile defenses, as recently discussed in several excellent reviews [17, 94, 95], will include the assessment of timing and synergy of volatile emissions with direct defense responses of the herbivore-damaged plant, especially if indirect defense strategies will be integrated with toxin-based pest control strategies.

Acknowledgments Our apologies to those research laboratories whose work we were unable to cite due to space constrictions. We thank Janet Webster and Jim Tokuhisa for helpful comments on the manuscript. Our work was supported by the US Department of Agriculture (NRI/CSREES-2007-35318-18384), the Thomas and Kate Jeffress Memorial Trust (J-850), and the Max Planck Society.

References

1. Pichersky E, Noel JP, Dudareva N (2006) Biosynthesis of plant volatiles: nature's diversity and ingenuity. Science 311:808–811
2. Goff SA, Klee HJ (2006) Plant volatile compounds: sensory cues for health and nutritional value. Science 311:815–819
3. Li JWH, Vederas JC (2009) Drug discovery and natural products: end of an era or an endless frontier. Science 325:161–165
4. Piechulla B, Pott MB (2003) Plant scents – mediators of inter- and intraorganismic communication. Planta 217:687–689
5. Rasmann S, Kollner TG, Degenhardt J, Hiltpold I, Toepfer S, Kuhlmann U, Gershenzon J, Turlings TCJ (2005) Recruitment of entomopathogenic nematodes by insect-damaged maize roots. Nature 434:732–737
6. Kappers IF, Aharoni A, van Herpen TWJM, Luckerhoff LLP, Dicke M, Bouwmeester HJ (2005) Genetic engineering of terpenoid metabolism attracts bodyguards to *Arabidopsis*. Science 309:2070–2072
7. Kessler D, Baldwin IT (2007) Making sense of nectar scents: the effects of nectar secondary metabolites on floral visitors of *Nicotiana attenuata*. Plant J 49:840–854
8. Clay NK, Adio AM, Denoux C, Jander G, Ausubel FM (2009) Glucosinolate metabolites required for an *Arabidopsis* innate immune response. Science 323:95–101
9. Bednarek P, Pislewska-Bednarek M, Svatos A, Schneider B, Doubsky J, Mansurova M, Humphry M, Consonni C, Panstruga R, Sanchez-Vallet A, Molina A, Schulze-Lefert P (2009) A glucosinolate metabolism pathway in living plant cells mediates broad-spectrum antifungal defense. Science 323:101–106
10. Schnee C, Koellner TG, Held M, Turlings TCJ, Gershenzon J, Degenhardt J (2006) The products of a single maize sesquiterpene synthase form a volatile defense signal that attracts natural enemies of maize herbivores. Proc Natl Acad Sci USA 103:1129–1134
11. Pichersky E, Gershenzon J (2002) The formation and function of plant volatiles: perfumes for pollinator attraction and defense. Curr Opin Plant Biol 5:237–243
12. Halkier BA, Gershenzon J (2006) Biology and biochemistry of glucosinolates. Annu Rev Plant Biol 57:303–333
13. Hartmann T (1999) Chemical ecology of pyrrolizidine alkaloids. Planta 207:483–495
14. Kessler A, Baldwin IT (2002) Plant responses to insect herbivory: the emerging molecular analysis. Annu Rev Plant Biol 53:299–328
15. Dicke M (2009) Behavioural and community ecology of plants that cry for help. Plant Cell Environ 32:654–665
16. Kessler A, Baldwin IT (2001) Defensive function of herbivore-induced plant volatile emissions in nature. Science 291:2141–2144

17. Turlings TCJ, Ton J (2006) Exploiting scents of distress: the prospect of manipulating herbivore-induced plant odours to enhance the control of agricultural pests. Curr Opin Plant Biol 9:421–427

18. Hilker M, Meiners T (2006) Early herbivore alert: insect eggs induce plant defense. J Chem Ecol 32:1379–1397

19. Poelman EH, Oduor AMO, Broekgaarden C, Hordijk CA, Jansen JJ, van Loon JJA, Van Dam NM, Vet LEM, Dicke M (2009) Field parasitism rates of caterpillars on *Brassica oleracea* plants are reliably predicted by differential attraction of *Cotesia* parasitoids. Funct Ecol 23:951–962

20. Turlings TCJ, Loughrin JH, McCall PJ, Rose USR, Lewis WJ, Tumlinson JH (1995) How caterpillar-damaged plants protect themselves by attracting parasitic wasps. Proc Natl Acad Sci USA 92:4169–4174

21. Yuan JS, Kollner TG, Wiggins G, Grant J, Degenhardt J, Chen F (2008) Molecular and genomic basis of volatile-mediated indirect defense against insects in rice. Plant J 55:491–503

22. Lou YG, Cheng JA (2003) Role of rice volatiles in the foraging behaviour of the predator *Cyrtorhinus lividipennis* for the rice brown planthopper *Nilaparvata lugens*. Biocontrol 48:73–86

23. Lou YG, Ma B, Cheng JA (2005) Attraction of the parasitoid *Anagrus nilaparvatae* to rice volatiles induced by the rice brown planthopper *Nilaparvata lugens*. J Chem Ecol 31:2357–2372

24. Kant MR, Ament K, Sabelis MW, Haring MA, Schuurink RC (2004) Differential timing of spider mite-induced direct and indirect defenses in tomato plants. Plant Physiol 135:483–495

25. Mercke P, Kappers IF, Verstappen FWA, Vorst O, Dicke M, Bouwmeester HJ (2004) Combined transcript and metabolite analysis reveals genes involved in spider mite induced volatile formation in cucumber plants. Plant Physiol 135:2012–2024

26. Gigolashvili T, Yatusevich R, Berger B, Mueller C, Fluegge UI (2007) The R2R3-MYB transcription factor HAG1/MYB28 is a regulator of methionine-derived glucosinolate biosynthesis in *Arabidopsis thaliana*. Plant J 51:247–261

27. Geu-Flores F, Olsen CE, Halkier BA (2009) Towards engineering glucosinolates into non-cruciferous plants. Planta 229:261–270

28. Khan ZR, AmpongNyarko K, Chiliswa P, Hassanali A, Kimani S, Lwande W, Overholt WA, Pickett JA, Smart LE, Wadhams LJ, Woodcock CM (1997) Intercropping increases parasitism of pests. Nature 388:631–632

29. Khan ZR, Pickett JA, Hassanali A, Hooper AM, Midega CAO (2008) *Desmodium* species and associated biochemical traits for controlling *Striga* species: present and future prospects. Weed Res 48:302–306

30. Beale MH, Birkett MA, Bruce TJA, Chamberlain K, Field LM, Huttly AK, Martin JL, Parker R, Phillips AL, Pickett JA, Prosser IM, Shewry PR, Smart LE, Wadhams LJ, Woodcock CM, Zhang YH (2006) Aphid alarm pheromone produced by transgenic plants affects aphid and parasitoid behavior. Proc Natl Acad Sci USA 103:10509–10513

31. Gouinguene S, Degen T, Turlings TCJ (2001) Variability in herbivore-induced odour emissions among maize cultivars and their wild ancestors (teosinte). Chemoecology 11:9–16

32. Turlings TCJ, Tumlinson JH, Lewis WJ (1990) Exploitation of herbivore-induced plant odors by host-seeking parasitic wasps. Science 250:1251–1253

33. Cheng AX, Xiang CY, Li JX, Yang CQ, Hu WL, Wang LJ, Lou YG, Chen XY (2007) The rice (*E*)-beta-caryophyllene synthase (OsTPS3) accounts for the major inducible volatile sesquiterpenes. Phytochemistry 68:1632–1641

34. Kollner TG, Held M, Lenk C, Hiltpold I, Turlings TCJ, Gershenzon J, Degenhardt J (2008) A maize (*E*)-beta-caryophyllene synthase implicated in indirect defense responses against herbivores is not expressed in most American maize varieties. Plant Cell 20:482–494

35. Degenhardt J, Hiltpold I, Kollner TG, Frey M, Gierl A, Gershenzon J, Hibbard BE, Ellersieck MR, Turlings TCJ (2009) Restoring a maize root signal that attracts insect-killing nematodes to control a major pest. Proc Natl Acad Sci USA 106:13213–13218

36. Boland W, Feng Z, Donath J, Gabler A (1992) Are acyclic C-11 and C-16 homoterpenes plant volatiles indicating herbivory. Naturwissenschaften 79:368–371

37. Ament K, Van Schie CC, Bouwmeester HJ, Haring MA, Schuurink RC (2006) Induction of a leaf specific geranylgeranyl pyrophosphate synthase and emission of (E,E)-4,8,12-trimethyltrideca-1,3,7,11-tetraene in tomato are dependent on both jasmonic acid and salicylic acid signaling pathways. Planta 224:1197–1208

38. Hopke J, Donath J, Blechert S, Boland W (1994) Herbivore-induced volatiles – the emission of acyclic homoterpenes from leaves of *Phaseolus lunatus* and *Zea mays* can be triggered by a α-glucosidase and jasmonic acid. FEBS Lett 352:146–150

39. Dicke M (1994) Local and systemic production of volatile herbivore-induced terpenoids: their role in plant-carnivore mutualism. J Plant Physiol 143:465–472

40. Dicke M, Vanbeek TA, Posthumus MA, Bendom N, Vanbokhoven H, Degroot AE (1990) Isolation and identification of volatile kairomone that affects acarine predator–prey interactions – involvement of host plant in its production. J Chem Ecol 16:381–396

41. Shimoda T, Ozawa R, Sano K, Yano E, Takabayashi J (2005) The involvement of volatile info-chemicals from spider mites and from food-plants in prey location of the generalist predatory mite *Neoseiulus californicus*. J Chem Ecol 31:2019–2032

42. de Boer JG, Posthumus MA, Dicke M (2004) Identification of volatiles that are used in discrimination between plants infested with prey or nonprey herbivores by a predatory mite. J Chem Ecol 30:2215–2230

43. Mumm R, Posthumus MA, Dicke M (2008) Significance of terpenoids in induced indirect plant defence against herbivorous arthropods. Plant Cell Environ 31:575–585

44. Aharoni A, Jongsma MA, Kim T-Y, Ri M-B, Giri AP, Verstappen FWA, Schwab W, Bouwmeester HJ (2006) Metabolic engineering of terpenoid biosynthesis in plants. Phytochem Rev 5:49–58

45. Bruce TJA, Matthes MC, Chamberlain K, Woodcock CM, Mohib A, Webster B, Smart LE, Birkett MA, Pickett JA, Napier JA (2008) cis-Jasmone induces *Arabidopsis* genes that affect the chemical ecology of multitrophic interactions with aphids and their parasitoids. Proc Natl Acad Sci USA 105:4553–4558

46. Arimura G-i, Ozawa R, Shimoda T, Nishioka T, Boland W, Takabayashi J (2000) Herbivory-induced volatiles elicit defence genes in lima bean leaves. Nature 406:512–515

47. Chen F, Tholl D, D'Auria JC, Farooq A, Pichersky E, Gershenzon J (2003) Biosynthesis and emission of terpenoid volatiles from *Arabidopsis* flowers. Plant Cell 15:481–494

48. Tholl D, Chen F, Petri J, Gershenzon J, Pichersky E (2005) Two sesquiterpene synthases are responsible for the complex mixture of sesquiterpenes emitted from *Arabidopsis* flowers. Plant J 42:757–771

49. Fäldt J, Arimura G, Gershenzon J, Takabayashi J, Bohlmann J (2003) Functional identification of AtTPS03 as (E)-β-ocimene synthase: a monoterpene synthase catalyzing jasmonate- and wound-induced volatile formation in *Arabidopsis thaliana*. Planta 216:745–751

50. Aharoni A, Giri AP, Deuerlein S, Griepink F, de Kogel WJ, Verstappen FWA, Verhoeven HA, Jongsma MA, Schwab W, Bouwmeester HJ (2003) Terpenoid metabolism in wild-type and transgenic *Arabidopsis* plants. Plant Cell 15:2866–2884

51. Tholl D (2006) Terpene synthases and the regulation, diversity and biological roles of terpene metabolism. Curr Opin Plant Biol 9:297–304

52. Aubourg S, Lecharny A, Bohlmann J (2002) Genomic analysis of the terpenoid synthase (*AtTPS*) gene family of *Arabidopsis thaliana*. Mol Genet Genomics 267:730–745

53. Tholl D, Chen F, Gershenzon J, Pichersky E, Romeo JT (2004) *Arabidopsis thaliana*, a model system for investigating volatile terpene biosynthesis, regulation, and function. In: Romeo JT (ed) Secondary metabolism in model systems. Elsevier, Amsterdam, pp 1–18

54. Chen F, Ro DK, Petri J, Gershenzon J, Bohlmann J, Pichersky E, Tholl D (2004) Characterization of a root-specific *Arabidopsis* terpene synthase responsible for the formation of the volatile monoterpene 1,8-cineole. Plant Physiol 135:1956–1966

55. Steeghs M, Bais HP, de Gouw J, Goldan P, Kuster W, Northway M, Fall R, Vivanco JM (2004) Proton-transfer-reaction mass spectrometry as a new tool for real time analysis of root-secreted volatile organic compounds in arabidopsis. Plant Physiol 135:47–58

56. Birnbaum K, Shasha DE, Wang JY, Jung JW, Lambert GM, Galbraith DW, Benfey PN (2003) A gene expression map of the *Arabidopsis* root. Science 302:1956–1960

57. Brady SM, Orlando DA, Lee JY, Wang JY, Koch J, Dinneny JR, Mace D, Ohler U, Benfey PN (2007) A high-resolution root spatiotemporal map reveals dominant expression patterns. Science 318:801–806

58. Herde M, Gartner K, Kollner TG, Fode B, Boland W, Gershenzon J, Gatz C, Tholl D (2008) Identification and regulation of *TPS04/GES*, an *Arabidopsis* geranyllinalool synthase catalyzing the first step in the formation of the insect-induced volatile C-16-homoterpene TMTT. Plant Cell 20:1152–1168

59. Van Poecke RMP, Posthumus MA, Dicke M (2001) Herbivore-induced volatile production by *Arabidopsis thaliana* leads to attraction of the parasitoid *Cotesia rubecula*: Chemical, behavioral, and gene-expression analysis. J Chem Ecol 27:1911–1928

60. van Loon JJA, de Boer JG, Dicke M (2000) Parasitoid-plant mutualism: parasitoid attack of herbivore increases plant reproduction. Entomol Exp Appl 97:219–227

61. Boland W, Gabler A, Gilbert M, Feng ZF (1998) Biosynthesis of C-11 and C-16 homoterpenes in higher plants – stereochemistry of the C-C-bond cleavage reaction. Tetrahedron 54: 14725–14736

62. Degenhardt J, Gershenzon J (2000) Demonstration and characterization of (*E*)-nerolidol synthase from maize: a herbivore-inducible terpene synthase participating in (3-*E*)-4, 8-dimethyl-1,3,7-nonatriene biosynthesis. Planta 210:815–822

63. Bouwmeester HJ, Verstappen FWA, Posthumus MA, Dicke M (1999) Spider mite-induced (3*S*)-(*E*)-nerolidol synthase activity in cucumber and lima bean. The first dedicated step in acyclic C11-homoterpene biosynthesis. Plant Physiol 121:173–180

64. Schnee C, Kollner TG, Gershenzon J, Degenhardt J (2002) The maize gene *terpene synthase 1* encodes a sesquiterpene synthase catalyzing the formation of (*E*)-β-farnesene, (*E*)-nerolidol, and (*E,E*)-farnesol after herbivore damage. Plant Physiol 130:2049–2060

65. Dudareva N, Cseke L, Blanc VM, Pichersky E (1996) Evolution of floral scent in *Clarkia*: Novel patterns of *S*-linalool synthase gene expression in the *C. breweri* flower. Plant Cell 8:1137–1148

66. Donath J, Boland W (1994) Biosynthesis of acyclic homoterpenes in higher plants parallels steroid hormone metabolism. J Plant Physiol 143:473–478

67. Akhtar M, Wright JN (1991) A unified mechanistic view of oxidative reactions catalyzed by P-450 and related Fe-containing enzymes. Nat Prod Rep 8:527–551

68. Larbat R, Kellner S, Specker S, Hehn A, Gontier E, Hans J, Bourgaud F, Matern U (2007) Molecular cloning and functional characterization of psoralen synthase, the first committed monooxygenase of furanocoumarin biosynthesis. J Biol Chem 282:542–554

69. Field B, Osbourn AE (2008) Metabolic diversification – Independent assembly of operon-like gene clusters in different plants. Science 320:543–547

70. Davis EM, Croteau R, Leeper FJ, Vederas JC (2000) Cyclization enzymes in the biosynthesis of monoterpenes, sesquiterpenes, and diterpenes. In: Topics in current chemistry: biosynthesis-aromatic polyketides, isoprenoids, alkaloids. Springer, Heidelberg, pp 53–95

71. Okada K, Saito T, Nakagawa T, Kawamukai M, Kamiya Y (2000) Five geranylgeranyl diphosphate synthases expressed in different organs are localized into three subcellular compartments in *Arabidopsis*. Plant Physiol 122:1045–1056

72. Gerber E, Hemmerlin A, Hartmann M, Heintz D, Hartmann MA, Mutterer J, Rodriguez-Concepcion M, Boronat A, Van Dorsselaer A, Rohmer M, Crowell DN, Bach TJ (2009) The plastidial 2-C-methyl-D-erythritol 4-phosphate pathway provides the isoprenyl moiety for protein geranylgeranylation in tobacco BY-2 cells. Plant Cell 21:285–300

73. Aharoni A, Giri AP, Verstappen FWA, Bertea CM, Sevenier R, Sun ZK, Jongsma MA, Schwab W, Bouwmeester HJ (2004) Gain and loss of fruit flavor compounds produced by wild and cultivated strawberry species. Plant Cell 16:3110–3131

74. Nagegowda DA, Gutensohn M, Wilkerson CG, Dudareva N (2008) Two nearly identical terpene synthases catalyze the formation of nerolidol and linalool in snapdragon flowers. Plant J 55:224–239

75. Turner GW, Croteau R (2004) Organization of monoterpene biosynthesis in *Mentha*. Immunocytochemical localizations of geranyl diphosphate synthase, limonene-6-hydroxylase, isopiperitenol dehydrogenase, and pulegone reductase. Plant Physiol 136:4215–4227

76. Arimura GI, Garms S, Maffei M, Bossi S, Schulze B, Leitner M, Mithoefer A, Boland W (2008) Herbivore-induced terpenoid emission in *Medicago truncatula*: concerted action of jasmonate, ethylene and calcium signaling. Planta 227:453–464

77. Lucker J, Bouwmeester HJ, Schwab W, Blaas J, van der Plas LHW, Verhoeven HA (2001) Expression of *Clarkia S*-linalool synthase in transgenic petunia plants results in the accumulation of *S*-linalyl-α-D-glucopyranoside. Plant J 27:315–324

78. Godard KA, Byun-McKay A, Levasseur C, Plant A, Seguin A, Bohlmann J (2007) Testing of a heterologous, wound- and insect-inducible promoter for functional genomics studies in conifer defense. Plant Cell Rep 26:2083–2090

79. Verdonk JC, Haring MA, van Tunen AJ, Schuurink RC (2005) ODORANT1 regulates fragrance biosynthesis in petunia flowers. Plant Cell 17:1612–1624

80. Gigolashvili T, Berger B, Mock HP, Muller C, Weisshaar B, Fluegge UI (2007) The transcription factor HIG1/MYB51 regulates indolic glucosinolate biosynthesis in *Arabidopsis thaliana*. Plant J 50:886–901

81. van der Fits L, Memelink J (2001) The jasmonate-inducible AP2/ERF-domain transcription factor ORCA3 activates gene expression via interaction with a jasmonate-responsive promoter element. Plant J 25:43–53

82. Naoumkina MA, He XZ, Dixon RA (2008) Elicitor-induced transcription factors for metabolic reprogramming of secondary metabolism in *Medicago truncatula*. BMC Plant Biol 8:132

83. Butelli E, Titta L, Giorgio M, Mock HP, Matros A, Peterek S, Schijlen EGWM, Hall RD, Bovy AG, Luo J, Martin C (2008) Enrichment of tomato fruit with health-promoting anthocyanins by expression of select transcription factors. Nat Biotechnol 26:1301–1308

84. Xu YH, Wang JW, Wang S, Wang JY, Chen XY (2004) Characterization of GaWRKY1, a cotton transcription factor that regulates the sesquiterpene synthase gene (+)-delta-cadinene synthase-A. Plant Physiol 135:507–515

85. Okada A, Okada K, Miyamoto K, Koga J, Shibuya N, Nojiri H, Yamane H (2009) OsTGAP1, a bZIP transcription factor, coordinately regulates the inductive production of diterpenoid phytoalexins in rice. J Biol Chem 284:26510–26518

86. Hirai MY, Sugiyama K, Sawada Y, Tohge T, Obayashi T, Suzuki A, Araki R, Sakurai N, Suzuki H, Aoki K, Goda H, Nishizawa OI, Shibata D, Saito K (2007) Omics-based identification of *Arabidopsis* Myb transcription factors regulating aliphatic glucosinolate biosynthesis. Proc Natl Acad Sci USA 104:6478–6483

87. Wallaart TE, Bouwmeester HJ, Hille J, Poppinga L, Maijers NCA (2001) Amorpha-4, 11-diene synthase: cloning and functional expression of a key enzyme in the biosynthetic pathway of the novel antimalarial drug artemisinin. Planta 212:460–465

88. Zook M, Hohn T, Bonnen A, Tsuji J, Hammerschmidt R (1996) Characterization of novel sesquiterpenoid biosynthesis in tobacco expressing a fungal sesquiterpene synthase. Plant Physiol 112:311–318

89. Wu SQ, Schalk M, Clark A, Miles RB, Coates R, Chappell J (2006) Redirection of cytosolic or plastidic isoprenoid precursors elevates terpene production in plants. Nat Biotechnol 24:1441–1447

90. Davidovich-Rikanati R, Lewinsohn E, Bar E, Iijima Y, Pichersky E, Sitrit Y (2008) Overexpression of the lemon basil alpha-zingiberene synthase gene increases both mono- and sesquiterpene contents in tomato fruit. Plant J 56:228–238
91. Aharoni A, Jongsma MA, Bouwmeester HJ (2005) Volatile science? Metabolic engineering of terpenoids in plants. Trends Plant Sci 10:594–602
92. Dudareva N, Pichersky E (2008) Metabolic engineering of plant volatiles. Curr Opin Biotechnol 19:181–189
93. Dicke M, van Loon JJA, Soler R (2009) Chemical complexity of volatiles from plants induced by multiple attack. Nat Chem Biol 5:317–324
94. Degenhardt J (2009) Indirect defense responses to herbivory in grasses. Plant Physiol 149: 96–102
95. Degenhardt J, Gershenzon J, Baldwin IT, Kessler A (2003) Attracting friends to feast on foes: engineering terpene emission to make crop plants more attractive to herbivore enemies. Curr Opin Biotechnol 14:169–176
96. Mithofer A, Wanner G, Boland W (2005) Effects of feeding *Spodoptera littoralis* on lima bean leaves. II. Continuous mechanical wounding resembling insect feeding is sufficient to elicit herbivory-related volatile emission. Plant Physiol 137:1160–1168

Chapter 12
Stereoselectivity of the Biosynthesis
of Norlignans and Related Compounds

Toshiaki Umezawa, Masaomi Yamamura, Tomoyuki Nakatsubo,
Shiro Suzuki, and Takefumi Hattori

12.1 Introduction

The functions of phenylpropanoid derivatives are as diverse as their structural variations. Phenylpropanoids serve as phytoalexins, UV protectants, insect repellents, flower pigments, and signal molecules for plant–microbe interactions. They also function as polymeric constituents of support and surface structures such as lignins and suberins [1]. Therefore, biosynthesis of phenylpropanoids has received much interest in relation to these functions. In addition, the biosynthesis of these compounds has been intensively studied because they are often chiral, and naturally occurring samples of these compounds are usually optically active. Elucidation of these enantioselective mechanisms may contribute to the development of novel biomimetic systems for enantioselective organic synthesis.

Lignin is biosynthesized via oxidative coupling of p-hydroxycinnamyl alcohols (monolignols) and related compounds that are formed in the cinnamate/monolignol pathway [2–7]. The coupling has not been shown to be controlled enantioselectively, and although each lignin molecule is chiral, the polymeric lignin, which is not an equimolar mixture of pairs of enantiomers, is not racemic [8]. Lignins consist of a number of substructures (dimeric or intermonomeric structures, e.g., β-*O*-4 and syringaresinol), each of which may in turn be composed of both enantiomers (e.g., the (+)-*erythro/threo*-β-*O*-4 and (−)-*erythro/threo*-β-*O*-4 substructures and the (+)-syringaresinol and (−)-syringaresinol substructures) [8, 9]. Thus, lignins may be racemate-like in terms of their substructures [8].

Lignans are phenylpropanoid dimers in which the monomers are linked by the central carbon (C8) atoms of the propyl side chains (Fig. 12.1) [10]. Many lignans are formed from coniferyl alcohol, a typical lignin monomer, and the coupling of two coniferyl alcohol radicals proceeds under the control of a unique asymmetric

T. Umezawa (✉)
Research Institute for Sustainable Humanosphere, Kyoto University, Kyoto 611-0011, Japan;
Institute of Sustainability Science, Kyoto University, Kyoto 611-0011, Japan
e-mail: tumezawa@rish.kyoto-u.ac.jp

D.R. Gang (ed.), *The Biological Activity of Phytochemicals*, Recent Advances
in Phytochemistry 41, DOI 10.1007/978-1-4419-7299-6_12,
© Springer Science+Business Media, LLC 2011

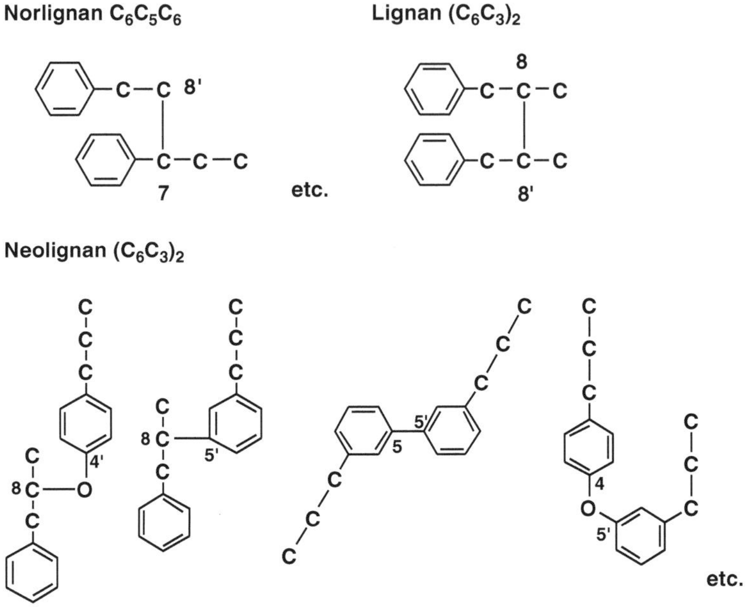

Fig. 12.1 Basic carbon skeletons of norlignans, lignans, and neolignans

inducer, dirigent protein (DP), giving rise to an optically active product, pinoresinol, which is further converted to a variety of lignans [11–13].

Neolignans are also compounds composed of two phenylpropanoid units, but linked in a manner other than C8-C8′ (Fig. 12.1) [10]. The compounds of this class are often chiral and naturally occurring neolignans are usually optically active, which evokes the involvement of DPs in the biosynthesis of neolignans.

Norlignans are a class of natural phenolic compounds with diphenylpentane carbon skeletons (C_6-C_5-C_6) (Fig. 12.1). A typical norlignan consists of a basic carbon skeleton containing C_6C_2 and C_6C_3 units linked via C7-C8′ bond. These intermonomer bonds cannot be formed by simple coupling of two phenoxyradicals formed from the phenylpropanoid monomers, suggesting the participation of an enantioselective mechanism(s) other than those assisted by DPs. Recently, Yamamura et al. [14] have found that the subunit compositions of hinokiresinol synthase can control the enantioselectivity in the formation of the norlignan, (Z)-hinokiresinol.

Herein, we briefly outline the recent advances in the studies of stereoselectivity in the biosynthesis of norlignans, lignans, and neolignans.

12.2 Norlignan Biosynthesis

Hinokiresinol is the simplest norlignan (Fig. 12.2) and (E)-hinokiresinol was first described by Hirose et al. [15]. Hinokiresinol has a C8-C7′-type carbon skeleton

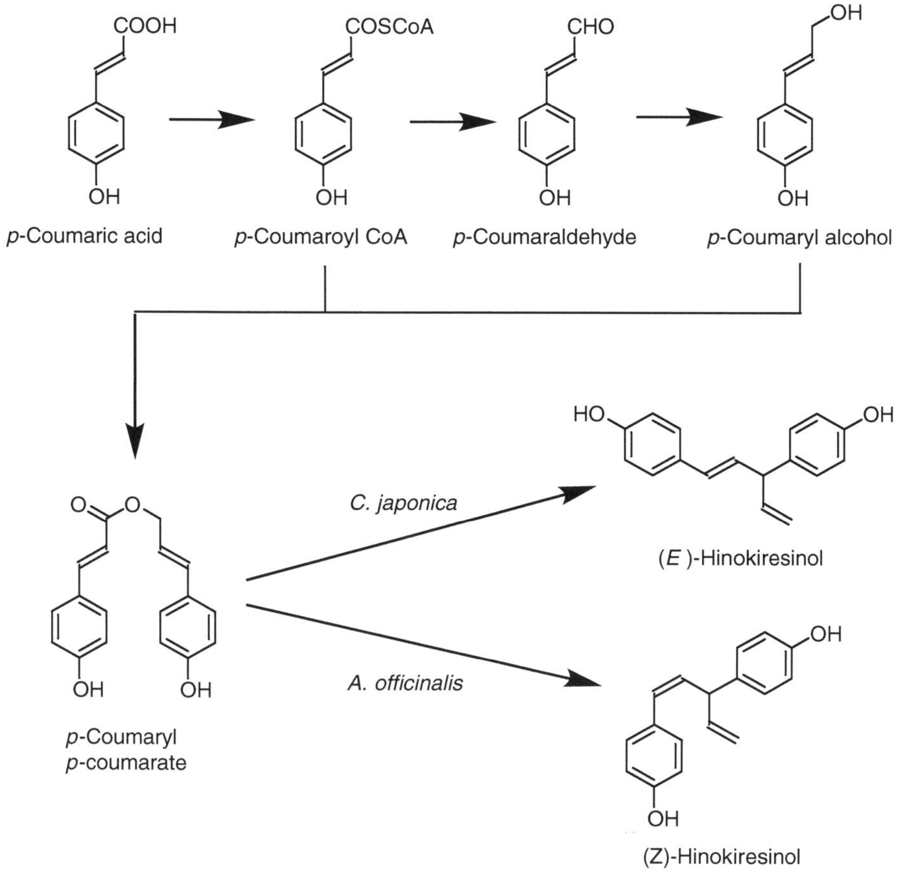

Fig. 12.2 Biosynthetic pathway of hinokiresinols

and many members of this norlignan class have been isolated from conifers and monocotyledonous plants [13]. In addition, C8-C8′- and C9-C8′-type norlignans are known [13].

Based on the chemical structures of the isolated norlignans, several hypothetical biosynthetic pathways were proposed [16–18]. All these proposed pathways involved coupling of two phenylpropanoid monomer units followed by a loss of one carbon atom to give rise to norlignans. Although this mechanism seemed plausible, another mechanism that involved the addition of two carbon atoms to flavonoid compounds (C_6-C_3-C_6) to give norlignans (C_6-C_5-C_6) could not be ruled out.

Later, a series of administration experiments with deuterium- and ^{13}C-labeled phenylpropanoid monomers and elicitor-treated, (Z)-hinokiresinol-forming cultured cells of *Asparagus officinalis* were carried out [19]. GC-MS and NMR analyses of the (Z)-hinokiresinol isolated following the administration indicated that all carbon atoms of (Z)-hinokiresinol were derived from phenylpropanoid monomers and that *p*-coumaryl alcohol and a *p*-coumaroyl compound, probably *p*-coumaroyl CoA, were likely immediate monomeric precursors of hinokiresinol formation (Fig. 12.2) [19].

The mechanism was confirmed by enzymatic experiments [20, 21]. A crude enzyme preparation from *A. officinalis* cultured cells catalyzed the conversion of *p*-coumaryl alcohol and *p*-coumaroyl CoA to (*Z*)-hinokiresinol [20], while a crude enzyme preparation from *Cryptomeria japonica* cultured cells mediated the formation of (*E*)-hinokiresinol from the same substrates [21]. In addition, both enzyme preparations converted *p*-coumaryl *p*-coumarate into (*Z*)-hinokiresinol and (*E*)-hinokiresinol [20, 21]. Thus, the biosynthesis of hinokiresinol originating from phenylpropanoid monomers was established.

The *A. officinalis* enzyme responsible for catalyzing (*Z*)-hinokiresinol formation, (*Z*)-hinokiresinol synthase (ZHRS), was purified and cDNAs encoding the enzyme were isolated [22]. When the enzyme was purified chromatographically, two proteins that showed slightly different masses in SDS-PAGE were obtained and designated ZHRSα (21 kDa) and ZHRSβ (23 kDa). Both ZHRSα and ZHRSβ were submitted to lysyl endpeptidase digestion followed by amino acid microsequencing. Based on the amino acid sequence information of the polypeptide fragments, cDNAs encoding the two proteins were obtained. Then, recombinant proteins of the two proteins, recZHRSα and recZHRSβ, were prepared and submitted for biochemical characterization. Surprisingly, when either recZHRSα or recZHRSβ was incubated individually with *p*-coumaryl *p*-coumarate, (*E*)-hinokiresinol, but not the (*Z*)-isomer that is the naturally occurring isomer in *A. officinalis*, was formed. On the other hand, the incubation of a mixture of equal parts of recZHRSα and recZHRSβ with the same substrate gave (*Z*)-hinokiresinol. Analysis by gel filtration chromatography showed that natural ZHRS, recZHRSα, and recZHRSβ were all 37–40 kDa. In addition, the protein migration patterns on a native PAGE gel were similar among recZHRSαβ, recZHRSβ, and the 1:1 mixture of recZHRSα and recZHRSβ, indicating that these proteins exist as dimers. Given that both recZHRSα and recZHRSβ are required for (*Z*)-hinokiresinol synthase activity, it is most likely that natural ZHRS exists as a heterodimer of ZHRSα and ZHRSβ, whereas recZHRSα and recZHRSβ are present as homodimers (recZHRSα×2 and recZHRSβ×2). Thus, ZHRSα and ZHRSβ are most likely to be subunits of ZHRS and subunit composition can control *cis/trans* isomerism of the product, hinokiresinol (Fig. 12.2) [22].

In natural product biosynthesis, the *E* and *Z* double bonds are usually formed selectively. For instance, *cis*-prenyltransferases catalyze the synthesis of (*Z*)-prenyl chains, while *trans*-prenyltransferases catalyze the formation of (*E*)-prenyl chains, although both use a common substrate, isopentenyl diphosphate [23]. In the biosynthesis of (*E*)-3,7-dimethyl-2,6-octadiene-1-ol (geraniol) and (*Z*)-3,7-dimethyl-2,6-octadiene-1-ol (nerol), geraniol is selectively formed from geranyl diphosphate by geraniol synthase [24], whereas the *Z*-isomer, nerol, is probably formed from geraniol via isomerization [24, 25]. Overall, the selective formation of *E* or *Z* double bonds is controlled by distinct enzymes. By contrast, the subunit compositions of ZHRS can control the *E* and *Z* selectivity. To the best of our knowledge, this is the first example of *cis/trans* double bond regulation by subunits of a single enzyme [22].

Based on the result that the ZHRS subunits can control *cis/trans* isomerism, it was also of interest to determine whether the subunit composition can control the enantiomeric selectivity during hinokiresinol formation. Both (*Z*)- and (*E*)-hinokiresinols have an asymmetric carbon atom at the C7 position. A literature survey indicated that the enantiomeric compositions of hinokiresinol can vary with the plant species; *Anemarrhena asphodeloides* (rhizome) gave (*Z*)-hinokiresinol in favor of (*R*)-enantiomer with 54% e.e. [26] and 36% e.e. [27] (Table 12.1), while (*E*)-hinokiresinol isolated from *Chamaecyparis obtusa* (heartwood) was suggested to be the optically pure (-)-(7*S*)-isomer [26] (Table 12.1). These results strongly suggest that hinokiresinol-producing plants have distinct mechanisms for controlling enantiomeric selectivity during hinokiresinol formation, as well as controlling *cis/trans* isomerism.

Although (*Z*)-hinokiresinol is stable during silica gel TLC separation, it was isomerized into the (*E*)-form after storage for 3 years in a freezer at −20°C (Yamamura, Suzuki, Hattori, Umezawa, unpublished). In addition, the (*E*)-isomer is rather unstable under slightly acidic conditions (such as old deutrochloroform for NMR measurements) (Umezawa, unpublished). Hence, we established a system for enantiomeric analysis of hinokiresinols by converting them into the corresponding

Table 12.1 Enantiomeric composition of (*Z*)-hinokiresinol and (*E*)-hinokiresinol

	Plant species	Enzyme	Predominant enantiomer	Enantiomeric composition (% e.e.)	References
Hinokiresinol formed following incubation with enzymes	–	Crude enzyme preparation from cell culture of *Asparagus officinalis*	(7*S*)-ZHR	97.2	[14]
	–	A mixture of recZHRSα and recZHRSβ (α:β=1:1)	(7*S*)-ZHR	100	[14]
	–	recZHRSα	(7*S*)-EHR	20.6	[14]
	–	recZHRSβ	(7*S*)-EHR	9.0	[14]
Hinokiresinol isolated from plants	*Asparagus officinalis* (cell cultures)	–	(7*S*)-ZHR	100	[14]
	Anemarrhena asphodeloides (stem)	–	(7*R*)-ZHR	36–54	[26, 27]
	Cryptomeria japonica (cell cultures)	–	(7*S*)-EHR	83.3	[14]
	Chamaecyparis obtusa (heartwood)	–	(7*S*)-EHR	100	[14, 26]

ZHR: (*Z*)-hinokiresinol; EHR: (*E*)-hinokiresinol.

tetrahydroderivative, tetrahydrohinokiresinol (THR), followed by chiral HPLC separation [14]. By hydrogenation, (−)-(7*S*)-(*E*)-hinokiresinol isolated from *C. obtusa* heartwood gave optically pure (+)-(7S)-THR, thereby it was established that (+)-(7*S*)-THR is derived from (7*S*)-(*Z*)-hinokiresinol and (7*S*)-(*E*)-hinokiresinol, while (−)-(7*R*)-THR is derived from (7*R*)-hinokiresinols.

Using this system, (*Z*)-hinokiresinol isolated from cultured cells of *A. officinalis* was determined to be the optically pure (7*S*)-isomer, while (*E*)-hinokiresinol isolated from cultured cells of *C. japonica* had 83.3% e.e. in favor of the (7*S*)-enantiomer (Table 12.1). The enzymatically formed (*Z*)-hinokiresinol obtained following incubation of *p*-coumaryl *p*-coumarate with a mixture of equal amounts of recZHRSα and recZHRSβ was found to be the optically pure (7*S*)-isomer, which is identical to that isolated from *A. officinalis* cells (Table 12.1). A similar result was obtained with the crude plant protein from *A. officinalis* cultured cells, where the formed (*Z*)-hinokiresinol was almost optically pure, 97.2% e.e. in favor of the (7*S*)-isomer (Table 12.1). In sharp contrast, when each subunit protein, recZHRSα or recZHRSβ, was individually incubated with *p*-coumaryl *p*-coumarate, (*E*)-hinokiresinol was formed (Table 12.1). The enantiomeric compositions of (*E*)-hinokiresinol thus formed were 20.6% e.e. (with recZHRSα) and 9.0% e.e. (with recZHRSβ) in favor of the (7*S*)-enantiomer (Table 12.1). Taken together, these results clearly indicate that the subunit composition of ZHRS controls not only *cis/trans* selectivity but also enantioselectivity in hinokiresinol formation (Fig. 12.3). This provides a novel example of enantiomeric control in the biosynthesis of natural products. Although the mechanism for the *cis/trans* selective and enantioselective reaction remains to be elucidated, for example by x-ray crystallography, the enantioselective mechanism totally differs from the enantioselectivity in biosynthesis of lignans, another class of phenylpropanoid compounds closely related to norlignans in terms of structure and biosynthesis.

12.3 Lignan Biosynthesis

Lignans can be classified into three categories: lignans with 9(9′)-oxygen, lignans without 9(9′)-oxygen, and dicarboxylic acid lignans (*p*-hydroxycinnamate dimers) [11]. Of the three categories, the study of the biosynthesis of lignans with 9(9′)-oxygen is the most advanced [11–13].

12.3.1 Lignans with 9(9′)-Oxygen

This type of lignan is generally formed by the enantioselective dimerization of two coniferyl alcohol units with the aid of DP to give rise to optically active pinoresinol (furofuran lignan) (Fig. 12.4) [6, 11–13, 28, 29]. Pinoresinol is then reduced by pinoresinol/lariciresinol reductase (PLR) via lariciresinol (furan lignan) to sec-oisolariciresinol (dibenzylbutane lignan) [6, 11–13, 30–42]. Secoisolariciresinol

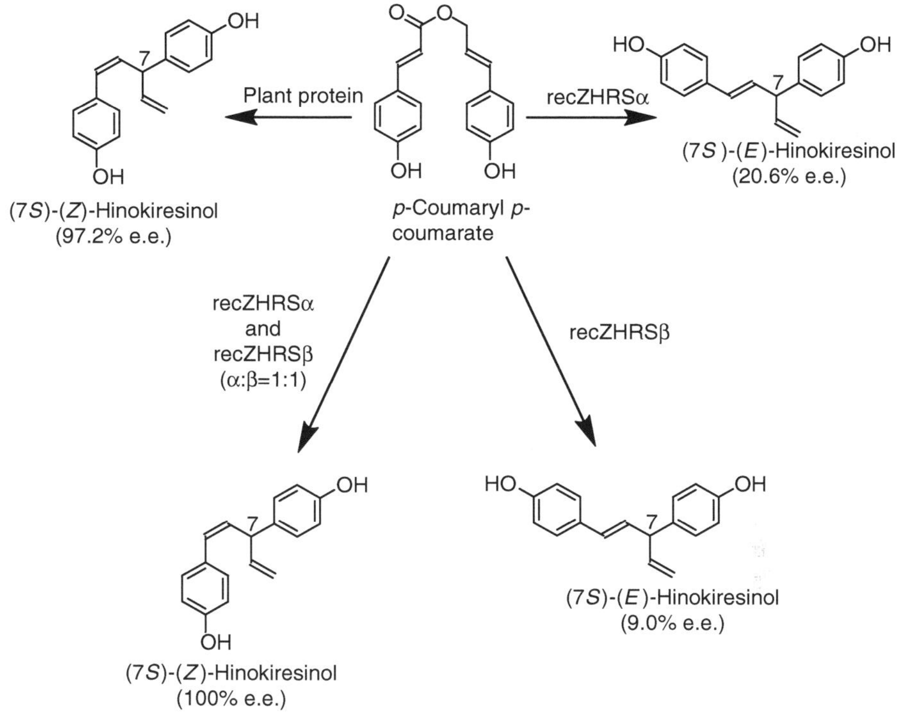

Fig. 12.3 Control of enantioselectivity in hinokiresinol formation by ZHRS subunits

is in turn oxidized to give matairesinol (dibenzylbutyrolactone lignan) by secoisolariciresinol dehydrogenase (SIRD) [6, 11–13, 43–48]. The conversion from coniferyl alcohol to matairesinol has been demonstrated in various plant species, which strongly suggests that this is the general lignan biosynthetic pathway. Many other subclasses of lignans are formed from the lignans found on the general pathway [11, 13].

For example, yatein and podophyllotoxin are biosynthesized from matairesinol (Fig. 12.4) [11, 13, 49]. This conversion involves 5-hydroxylation of one of the 3-methoxy-4-hydroxylphenyl (guaiacyl) moieties of matairesinol to give rise to 5-hydroxymatairesinol (thujaplicatin) followed by dual methylation at the 4- and 5-hydroxyls, giving rise to a 3,4,5-trialkoxyphenyl moiety (Fig. 12.4) [11, 13, 49].

On the other hand, biosynthetic pathways that do not involve the conversion from coniferyl alcohol to matairesinol have been proposed for lignans composed of two syringyl (3,5-dimethoxy-4-hydroxyphenyl) groups: (+)-syringaresinol formation in *Liriodendron tulipifera* [50] and (+)-lyoniresinol biosynthesis in *Lyonia ovalifolia* var. *elliptica* [51]. Enantioselective coupling of two sinapyl alcohol units was proposed for the selective formation of (+)-syringaresinol [50]. On the other hand, a non-enantioselective dimerization of sinapyl alcohol was proposed for (+)-lyoniresinol biosynthesis and the enantioselectivity in the biosynthesis was ascribed

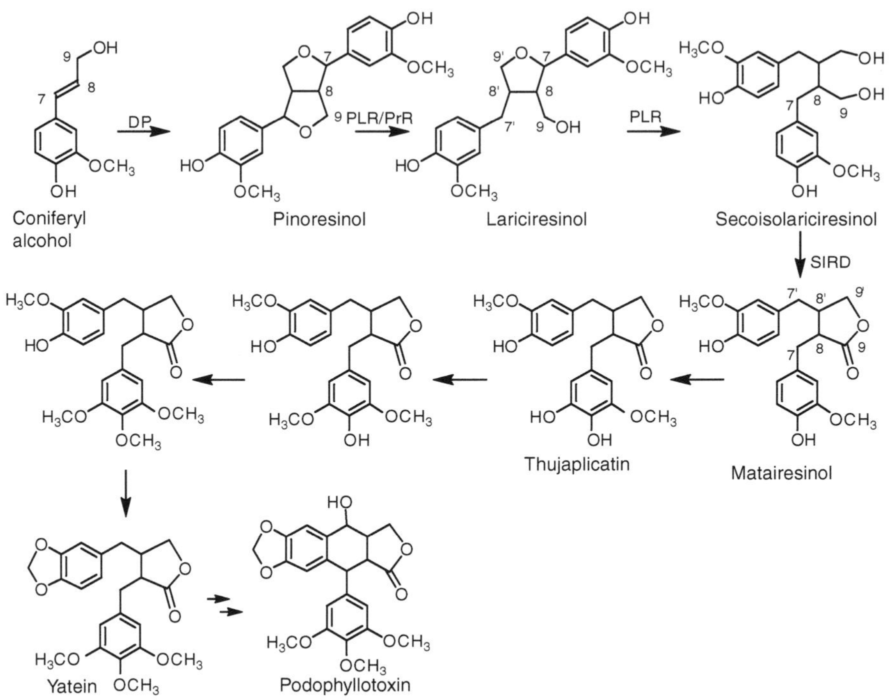

Fig. 12.4 Biosynthetic pathway for lignans with 9,9′-oxygen including yatein and podophyllotoxin. DP: dirigent protein, PLR: pinoresinol/lariciresinol reductase, PrR: pinoresinol reductase, and SIRD: secoisolariciresinol dehydrogenase

mainly to the formation of (+)-lyoniresinol from 5,5′-dimethoxylariciresinol (Fig. 12.5) [51].

DP is a unique and interesting asymmetric inducer, which controls central chirality affording optically active pinoresinol that has chiral centers at C8 and C8′, the positions participating in the intermonomer linkage [6, 12, 28, 29]. *Forsythia intermedia* DP mediates the enantioselective formation of (+)-pinoresinol from coniferyl alcohol radical (Fig. 12.6) [6, 12, 28, 29]. In addition, a DP-like protein that controls axial chirality has been reported [52]. This DP-like protein controlled the atropisomerism and aided the stereoselective coupling of an achiral terpenoid phenol, hemigossypol, giving rise to an optically active dimer, (+)-gossypol with 30% atropisomeric excess [52]. This may suggest the wide distribution of enantioselective DPs in plant secondary metabolism.

However, the enantiomeric control by DPs does not lead to the production of optically pure pinoresinol in plants, because the enantiomeric compositions of pinoresinol from various plant species vary widely and optically pure pinoresinol has not yet been isolated from plants [11, 13, 53]. Downstream lignans in the lignan biosynthetic pathway, such as dibenzylbutyrolactone lignans including matairesinol, are optically pure [11, 13, 53]. These facts unequivocally indicate that not only was

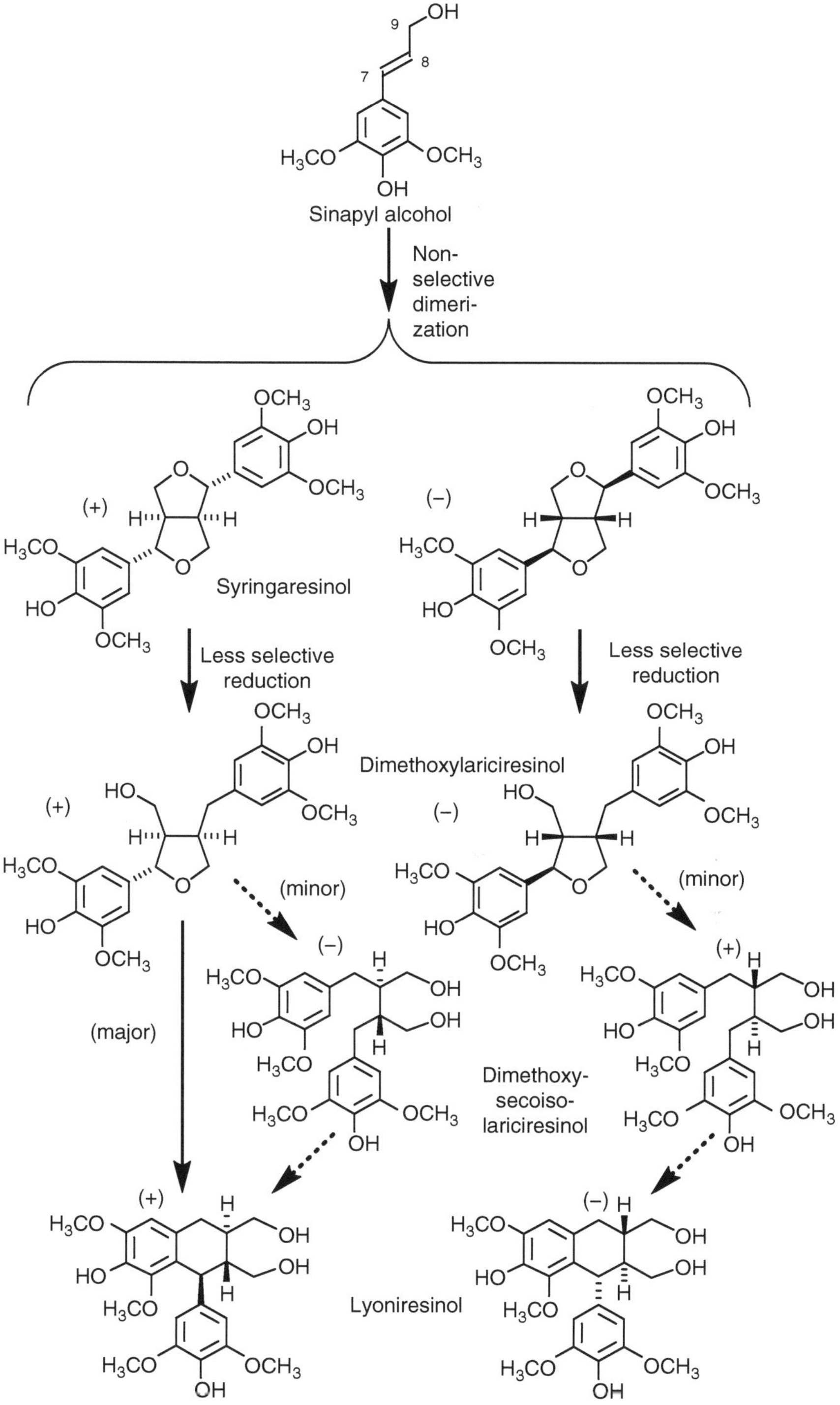

Fig. 12.5 A proposed biosynthetic pathway for (+)-lyoniresinol. (+):(+)-enantiomer, (−):(−)-enantiomer

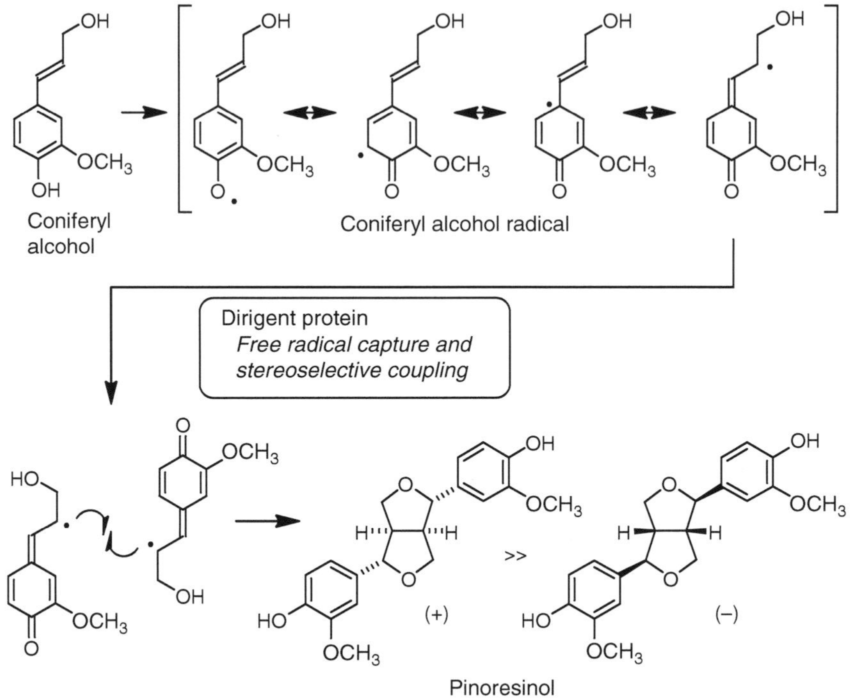

Fig. 12.6 Enantioselective formation of pinoresinol. *Forsythia intermedia* dirigent protein mediates selective formation of (+)-pinoresinol

the first step mediated by DP, but the subsequent metabolic steps catalyzed by PLR and SIRD were also involved in determining the enantiomeric composition of the lignans [54]. It has also been suggested that a variation in the enantiomeric composition of the upstream lignans (lariciresinol, secoisolariciresinol, and matairesinol) among different plant species may be ascribed, at least in part, to the selectivities of PLR and SIRD isoforms in terms of substrate enantiomers as well as their spatiotemporal expression patterns [54].

Recently, this view has been proven in relation to gene expression. Thus, in the *A. thaliana* genome database, there are two genes (At1g32100 and At4g13660) that are annotated as *PLR*. The recombinant AtPLRs showed strict substrate preference toward pinoresinol but only weak or no activity toward lariciresinol [55], which is in sharp contrast to conventional PLRs of other plants that can reduce both pinoresinol and lariciresinol efficiently to lariciresinol and secoisolariciresinol, respectively (Fig. 12.7) [34, 37, 41, 42]. Therefore, AtPLRs were renamed *A. thaliana* pinoresinol reductases (AtPrRs) [55]. The recombinant AtPrR2 encoded by At4g13660 reduced only (−)-pinoresinol to (−)-lariciresinol and not (+)-pinoresinol in the presence of NADPH. This enantiomeric selectivity is in accordance with that of PLRs of other plants so far reported, which can reduce one of the enantiomers selectively, whatever the preferential enantiomer [55]. In sharp contrast, AtPrR1 encoded by At1g32100 reduced both (+)-pinoresinol

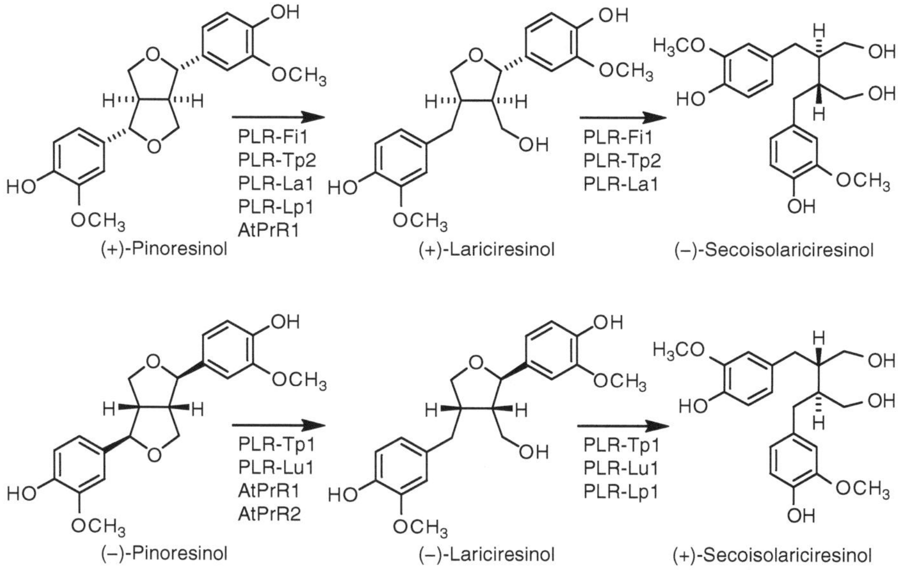

Fig. 12.7 Reactions catalyzed by pinoresinol/lariciresinol reductases from *Forsythia intermedia* (PLR-Fi1) [34], *Thuja plicata* (PLR-Tp1 and PLR-Tp2) [37], *Linum album* (PLR-La1) [41], *Linum perenne* (PLR-Lp1) [42], and *Linum usitatissimum* (PLR-Lu1) [41] and pinoresinol reductases from *Arabidopsis thaliana* (AtPrR1 and AtPrR2) [55]

and (−)-pinoresinol to (+)- and (−)-lariciresinols efficiently with comparative k_{cat}/K_m values.

Next, the substrate enantiomer selectivity obtained from in vitro experiments was compared with enantiomeric compositions of lariciresinol occurring in *A. thaliana* plants to examine the roles of *AtPrR1* and *AtPrR2* in the plant. Thus, homozygous lines of T-DNA insertion mutants of *AtPrR1* and *AtPrR2* genes and their double knockout mutant were prepared. GC-MS analysis indicated that the lariciresinol content in the single mutants, *atprr1-1*, *atprr1-2*, and *atprr2*, was not reduced, but was slightly increased compared with the wild type. Lariciresinol was not detected in the *atprr1-1 atprr2* double mutant. Instead, significant amounts of pinoresinol were detected in the double mutant. These results clearly indicate the functional redundancy of the two *AtPrR* genes in lariciresinol biosynthesis [55].

The enantiomeric composition of lariciresinol occurring in the wild type was 88% e.e. in favor of the (−)-enantiomer. Lariciresinol isolated from the *atprr1-1* and *atprr1-2* single mutants showed 94 and 96% e.e. in favor of the (−)-enantiomer, which are higher than that isolated from the wild type. This agrees well with the fact that AtPrR1 reduces both (+)-pinoresinol and (−)-pinoresinol almost equally. In the mutants, the deficiency of the isoform was compensated for by the AtPrR2 that preferentially reduces (−)-pinoresinol, giving rise to (−)-lariciresinol, which probably resulted in the increase of the percentage e.e. value in favor of the (−)-enantiomer. On the other hand, lariciresinol from the *atprr2* mutant had an 82% e.e. in favor of the (−)-enantiomer, which is lower than that of the wild type. Again, this can be accounted for by considering the fact that AtPrR2 reduces (−)-pinoresinol selectively. The deficiency of the isoform was compensated for by AtPrR1 that reduces both (+)-pinoresinol and (−)-pinoresinol almost equally, most probably resulting in the decrease of the percentage e.e. value compared with the wild type (88% e.e.). Although AtPrR1 reduced racemic (±)-pinoresinols to afford almost racemic lariciresinol (6% e.e. in favor of the (−)-enantiomer), the percentage e.e. value of the formed lariciresinol is much higher than the racemate. This can be ascribed to the enantioselective pinoresinol formation mediated by the DP. In fact, pinoresinol isolated from the *atprr1-1 atprr2* double mutant had 74% e.e. in favor of the (–)-pinoresinol. As long as (−)-pinoresinol with a high percentage e.e. value is supplied, less enantioselective AtPrR1 can produce (−)-lariciresinol with a high percentage e.e. value, as observed in the case of the *atprr2* mutant [55].

Taken together, it was demonstrated conclusively that differential expression of PrR isoforms that have distinct selectivities of substrate enantiomers plays a significant role in determining enantiomeric compositions of the product, lariciresinol, in addition to DP [55].

12.3.2 Other Types of Lignans

In 1972, Gottlieb [56] suggested that lignans and neolignans without 9(9′)-oxygen may be formed by coupling propenylphenols such as isoeugenol. This was supported by the similarity between the phylogenetic distribution of plants producing lignans without 9(9′)-oxygen and that of phenylpropanoid monomers without 9-oxygen, that is, allyl- and propenylphenols [11]. Furthermore, coupling of *p*-anol (*p*-propenylphenol) to give several furan lignans without 9(9′)-oxygen in *Larrea tridentata* was proposed [57]. Besides, administration of isoeugenol resulted in its conversion to a furan lignan without 9(9′)-oxygen, verrucosin, in *Virola surinamensis* (Fig. 12.8) [58]. Recently, when the radioisotope-labeled phenylpropanoid monomers, [U-^{14}C]phenylalanine, [9-^{3}H$_1$]coniferyl alcohol, and [9-^{3}H$_1$]isoeugenol, were administered to *Holostylis reniformis*, the incorporation of ^{3}H and ^{14}C into 10 aryltetralone lignans and 2 furan lignans was observed [59]. These experiments provide evidence that isoeugenol is a biosynthetic intermediate to the aryltetralone and furan lignans without 9(9′)-oxygen.

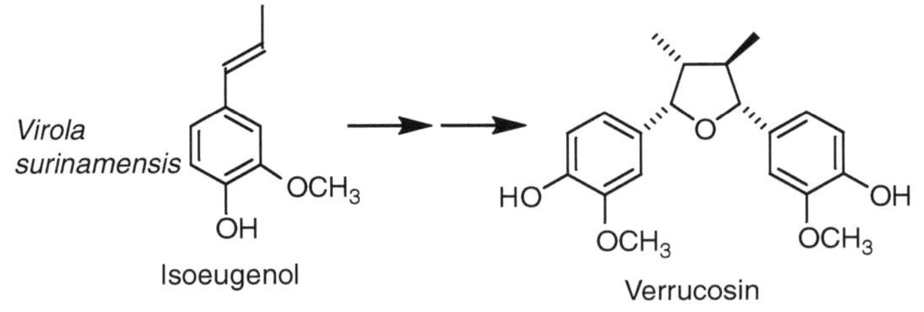

Fig. 12.8 Biosynthesis of a lignan without 9,9′-oxygen, verrucosin from isoeugenol

Despite frequent occurrence of allyl- and propenylphenols in plants, their biosynthetic mechanisms have only recently been elucidated. An NAD(P)H-dependent reductase in *Ocimum basilicum* (sweet basil) catalyzed the conversion of *p*-coumaryl *p*-coumarate and *p*-coumaryl acetate to chavicol (*p*-allylphenol) [60]. In addition, genes encoding NADPH-dependent reductases were isolated from *Petunia hybrida* cv. Mitchell (petunia) and *O. basilicum* [61]. The *Petunia* and *Ocimum* enzymes that catalyzed the conversion of coniferyl acetate to isoeugenol and eugenol were named PhIGS1 (isoeugenol synthase 1) and ObEGS1 (eugenol synthase 1), respectively [61]. Recently, a cDNA encoding an NADPH-dependent enzyme was obtained from *Pimpinella anisum* (anise) [62]. This enzyme, *t*-anol/isoeugenol synthase 1 (AIS1), can synthesize *t*-anol and isoeugenol from coumaryl acetate and coniferyl acetate, respectively [62]. In addition, a gene encoding an acetyltransferase that was effective with coniferyl alcohol was isolated [63]. Thus, the biosynthetic pathway from *p*-hydroxycinnamyl alcohols to allyl- or propenylphenols via esters of *p*-hydroxycinnamyl alcohols has been established.

Lignans of another type, dicarboxylic acid lignans (*p*-hydroxycinnamate dimers), were isolated from liverworts (Hepaticopsida) and some vascular plants [11, 64]. Little is known about the biosynthesis of this class of lignans, but a feeding experiment demonstrated the conversion of caffeic acid to optically pure epiphyllic acid in a liverwort *Lophocolea heterophylla* (Fig. 12.9) [65].

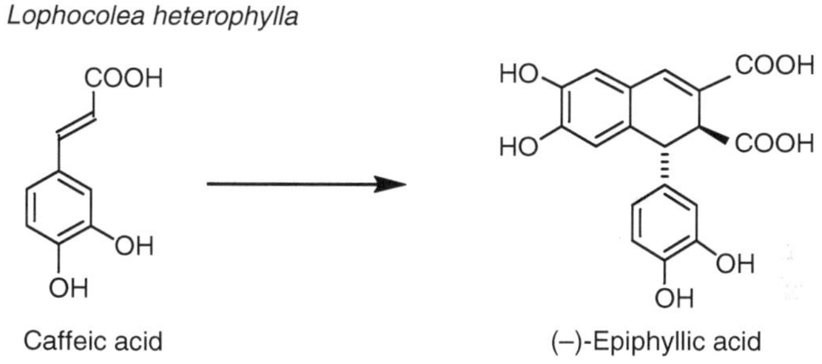

Fig. 12.9 Biosynthesis of a dicarboxylic acid lignan, epiphyllic acid

The biosynthetic pathways toward the three subclasses of lignans, lignans with 9(9′)-oxygen, lignans without 9(9′)-oxygen, and dicarboxylic acid lignans, branch off at the monomer level. DPs that function in the phenylpropanoid monomer coupling toward lignans without 9(9′)-oxygen and dicarboxylic acid lignans remains to be elucidated.

12.3.3 Biosynthesis of Neolignans

Neolignans are phenylpropanoid dimers in which the phenylpropanoid monomers are linked in a manner other than the lignan type, C8-C8′ bond (Fig. 12.1) [10]. Like

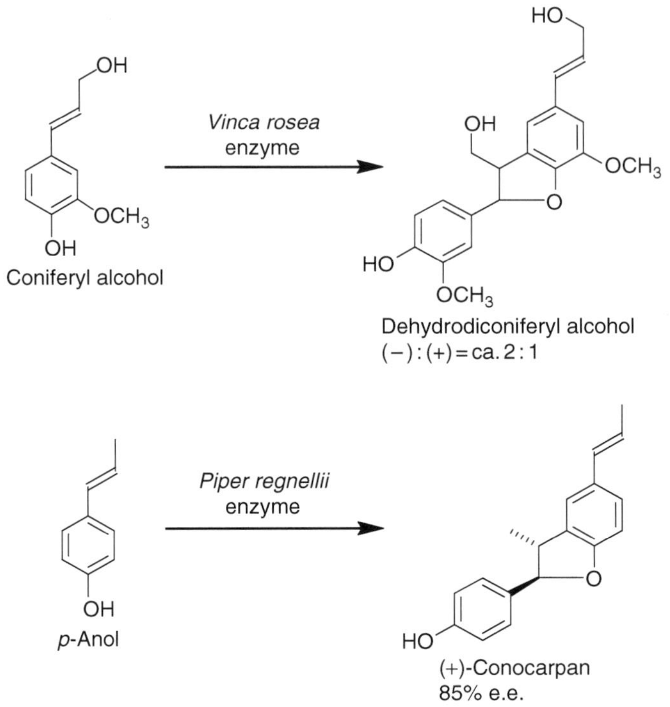

Fig. 12.10 Enantioselective formation of neolignans in *Vinca rosea* and *Piper regnellii*

the C8-C8′ bonds in lignans, intermonomer linkages of neolignans, such as C8-O-C4′ and C5-C8′, can be formed by coupling of two phenoxyradical units derived from phenylpropanoid monomers. The fact that naturally occurring neolignans are generally optically active strongly suggests that neolignan biosynthesis is aided by asymmetric inducers, similar to DP for lignan biosynthesis. Although neolignan DPs have not yet been identified, several studies have reported the enzymatic enantioselective formation of neolignans (Fig. 12.10).

A crude enzyme preparation from *Vinca rosea* catalyzed the enantioselective coupling of coniferyl alcohol to give rise to an optically active C8-C5′ neolignan, dehydrodiconiferyl alcohol, with a ratio of (−)-enantiomer to (+)-enantiomer of ~2:1 (Fig. 12.10) [66].

Another example of the enzymatic enantioselective formation of a C8-C5′ neolignan was reported. A crude enzyme preparation from *Piper regnellii* catalyzed the enantioselective formation of (+)-conocarpan (85% e.e.) from *p*-anol (Fig. 12.10) [67].

Incubation of coniferyl alcohol with a crude enzyme preparation from *Eucommia ulmoides* in the presence of hydrogen peroxide resulted in optically active C8-O-C4′ neolignan dimers, (+)-*erythro*-guaiacylglycerol-β-coniferyl ether (38% e.e.), and (-)-*threo*-guaiacylglycerol-β-coniferyl ether (34% e.e.) (Fig. 12.11) [68]. In

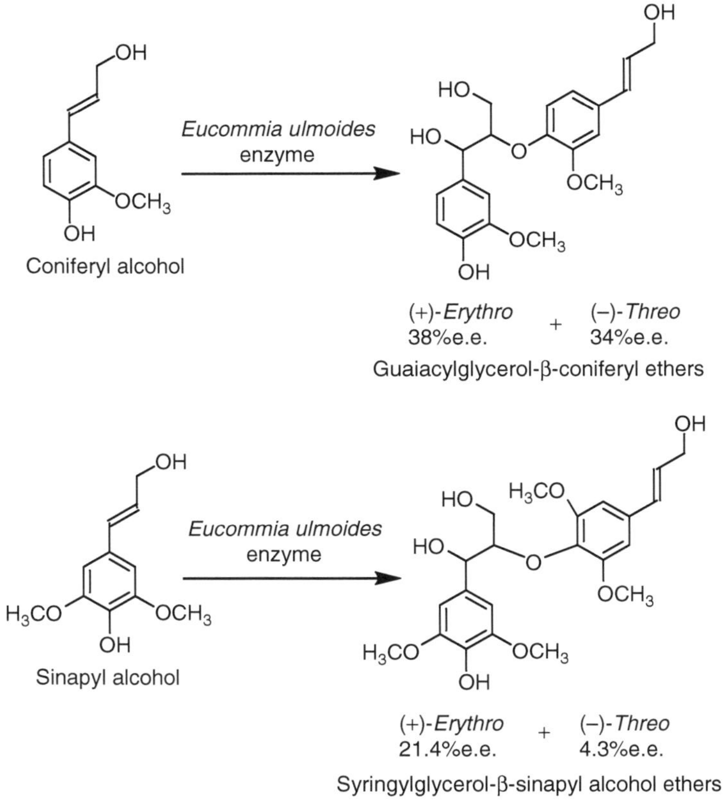

Fig. 12.11   Enantioselective formation of neolignans in *Eucommia ulmoides*

addition, an insoluble enzyme preparation from *E. ulmoides* catalyzed the enantioselective coupling of sinapyl alcohol to afford the optically active C8-O-C4′-type neolignans, (+)-*erythro*-syringylglycerol-β-sinapyl alcohol ether (21.4% e.e.), and (–)-*threo*-syringylglycerol-β-sinapyl alcohol ether (4.3% e.e.) (Fig. 12.11) [69]. These studies strongly suggested the involvement of asymmetric inducers in neolignan biosynthesis, DPs for neolignans. However, their identification and gene cloning await further experiments.

12.4 Concluding Remarks

A norlignan synthase, *A. officinalis* (Z)-hinokiresinol synthase (ZHRS), is composed of two subunits, ZHRSα and ZHRSβ. A 1:1 mixture of both subunits forms optically pure (Z)-hinokiresinol from *p*-coumaryl *p*-coumarate. By contrast, individual incubation of each subunit with the same substrate affords (E)-hinokiresinol with only

9.0–20.6% e.e. These results indicated that the subunit composition of ZHRS controls both *cis/trans* selectivity and enantioselectivity in hinokiresinol formation. This is a novel example of enantiomeric control in the biosynthesis of natural products. This mechanism differs from the enantioselective coupling of two coniferyl alcohol units in the biosynthesis of lignans with 9(9′)-oxygen, where dirigent proteins control enantioselectivity. In addition, pinoresinol (/lariciresinol) reductases involved in the subsequent lignan biosynthetic steps also play significant roles in the enantioselective formation of lignans. Formation of lignans of other types, lignans without 9(9′)-oxygen and dicarboxylic acid lignans, as well as biosynthesis of neolignans are also suggested to proceed with the aid of other types of dirigent proteins, but their identification awaits further experiments.

References

1. Hahlbrock K, Scheel D (1989) Physiology and molecular biology of phenylpropanoid metabolism. Annu Rev Plant Physiol Plant Mol Biol 40:347–369
2. Boerjan W, Ralph J, Baucher M (2003) Lignin biosynthesis. Annu Rev Plant Biol 54:519–546
3. Dixon RA, Reddy MSS (2003) Biosynthesis of monolignols. Genomic and reverse genetic approaches. Phytochem Rev 2:289–306
4. Ralph J, Lundquist K, Brunow G et al (2004) Lignins: natural polymers from oxidative coupling of 4-hydroxyphenylpropanoids. Phytochem Rev 3:29–60
5. Chiang VL (2006) Monolignol biosynthesis and genetic engineering of lignin in trees, a review. Environ Chem Lett 4:143–146
6. Davin LB, Jourdes M, Patten AM et al (2008) Dissection of lignin macromolecular configuration and assembly: comparison to related biochemical processes in allyl/propenyl phenol and lignan biosynthesis. Nat Prod Rep 25:1015–1090
7. Umezawa T (2010) The cinnamate/monolignol pathway. Phytochem Rev 9:1–17
8. Umezawa T (1997) Lignans. In: Higuchi T (ed) Springer series in wood science, biochemistry and molecular biology of wood. Springer, Berlin, pp. 181–194
9. Akiyama T, Magara K, Matsumoto Y et al (2000) Proof of the presence of racemic forms of arylglycerol-β-aryl ether structure in lignin: studies on the stereo structure of lignin by ozonation. J Wood Sci 46:414–415
10. Moss GP (2000) Nomenclature of lignans and neolignans. Pure Appl Chem 72:1493–1523
11. Umezawa T (2003) Diversity in lignan biosynthesis. Phytochem Rev 2:371–390
12. Davin L, Lewis NG (2003) An historical perspective on lignan biosynthesis: monolignol, allylphenol and hydroxycinnamic acid coupling and downstream metabolism. Phytochem Rev 2:257–288
13. Suzuki S, Umezawa T (2007) Biosynthesis of lignans and norlignans. J Wood Sci 53: 273–284
14. Yamamura M, Suzuki S, Hattori T et al (2010) Subunit composition of hinokiresinol synthase controls enantiomeric selectivity in hinokiresinol formation. Org Biomol Chem 8:1106–1110
15. Hirose Y, Oishi N, Nagaki H et al (1965) The structure of hinokiresinol. Tetrahedron Lett 41:3665–3668
16. Beracierta AP, Whiting DA (1978) Stereoselective total syntheses of the (±)-di-*O*-methyl ethers of agatharesinol, sequirin-A, and hinokiresinol, and of (±)-tri-*O*-methylsequrin-E, characteristic norlignans of Coniferae. J Chem Soc Perkin Trans 1:1257–1263
17. Birch AJ, Liepa AJ (1978) Biosynthesis. In: Rao CBS (ed) Chemistry of lignans. Andhra University Press, Andhra Pradesh, pp 307–327
18. Erdtman H, Harmatha J (1979) Phenolic and terpenoid heartwood constituents of *Libocedrus yateensis*. Phytochemistry 18:1495–1500

19. Suzuki S, Umezawa T, Shimada M (2001) Norlignan biosynthesis in *Asparagus officinalis* L.: the norlignan originates from two nonidentical phenylpropane units. J Chem Soc Perkin Trans 1:3252–3257

20. Suzuki S, Nakatsubo T, Umezawa T et al (2002) First in vitro norlignan formation with *Asparagus officinalis* enzyme preparation. Chem Commun 10:1088–1089

21. Suzuki S, Yamamura M, Shimada M et al (2004) A heartwood norlignan, (*E*)-hinokiresinol, is formed from 4-coumaryl 4-coumarate by a *Cryptomeria japonica* enzyme preparation. Chem Commun:2838–2839

22. Suzuki S, Yamamura M, Hattori T et al (2007) The subunit composition of hinokiresinol synthase controls geometrical selectivity in norlignan formation. Proc Natl Acad Sci USA 104:21008–21013

23. Kahrel Y, Takahashi S, Yamashita S et al (2006) Manipulation of prenyl chain length determination mechanism of *cis*-prenyltransferases. FEBS J 273:647–657

24. Iijima Y, Gang DR, Fridman E et al (2004) Characterization of geraniol synthase from the peltate glands of sweet basil. Plant Physiol 134:370–379

25. Wise ML, Croteau R (1999) Monoterpene biosynthesis. In: Barton D, Nakanishi K, Meth-Cohn O (eds) Comprehensive natural products chemistry, vol. 2. Elsevier Science, Oxford, pp 97–153

26. Minami E, Taki M, Takaishi S et al (2000) Stereochemistry of *cis*- and *trans*-hinokiresinol and their estrogen-like activity. Chem Pharm Bull 48:389–392

27. Jeong SJ, Higuchi R, Ono M et al (2003) *cis*-Hinokiresinol, a norlignan from *Anemarrhene asphodeloides*, inhibits angiogenic response in vitro and in vivo. Biol Pharm Bull 26: 1721–1724

28. Davin LB, Wang H-B, Crowell AL et al (1997) Stereoselective bimolecular phenoxy radical coupling by an auxiliary (dirigent) protein without an active center. Science 275:362–366

29. Gang DR, Costa MA, Fujita M et al (1999) Regiochemical control of monolignol radical coupling: a new paradigm for lignin and lignan biosynthesis. Chem Biol 6:143–151

30. Katayama T, Davin LB, Lewis NG (1992) An extraordinary accumulation of (-)-pinoresinol in cell-free extracts of *Forsythia intermedia*: evidence for enantiospecific reduction of (+)-pinoresinol. Phytochemistry 31:3875–3881

31. Chu A, Dinkova A, Davin LB et al (1993) Stereospecificity of (+)-pinoresinol and (+)-lariciresinol reductases from *Forsythia intermedia*. J Biol Chem 268:27026–27033

32. Umezawa T, Kuroda H, Isohata T et al (1994) Enantioselective lignan synthesis by cell-free extracts of *Forsythia koreana*. Biosci Biotech Biochem 58:230–234

33. Umezawa T, Shimada M (1996) Formation of the lignan (+)-secoisolariciresinol by cell-free extracts of *Arctium lappa*. Biosci Biotech Biochem 60:736–737

34. Dinkova-Kostova AT, Gang DR, Davin LB et al (1996) (+)-Pinoresinol/(+)-lariciresinol reductase from *Forsythia intermedia*. J Biol Chem 271:29473–29482

35. Katayama T, Masaoka T, Yamada H (1997) Biosynthesis and stereochemistry of lignans in *Zanthoxylum ailanthoides* I. (+)-Lariciresinol formation by enzymatic reduction of (±)-pinoresinols. Mokuzai Gakkaishi 43:580–588

36. Suzuki S, Umezawa T, Shimada M (1998) Stereochemical difference in secoisolariciresinol formation between cell-free extracts from petioles and from ripening seeds of *Arctium lappa* L. Biosci Biotech Biochem 62:1468–1470

37. Fujita M, Gang DR, Davin LB et al (1999) Recombinant pinoresinol-lariciresinol reductases from western red cedar (*Thuja plicata*) catalyze opposite enantiospecific conversions. J Biol Chem 274:618–627

38. Xia Z-Q, Costa MA, Proctor J et al (2000) Dirigent-mediated podophyllotoxin biosynthesis in *Linum flavum* and *Podophyllum peltatum*. Phytochemistry 55:537–549

39. Okunishi T, Umezawa T, Shimada M (2001) Isolation and enzymatic formation of lignans of *Daphne genkwa* and *Daphne odora*. J Wood Sci 47:383–388

40. Min T, Kasahara H, Bedgar DL et al (2003) Crystal structures of pinoresinol-lariciresinol and phenylcoumaran benzylic ether reductases and their relationship to isoflavone reductases. J Biol Chem 278:50714–50723

41. von Heimendahl CBI, Schäfer KM, Eklund P et al (2005) Pinoresinol-lariciresinol reductases with different stereospecificity from *Linum album* and *Linum usitatissimum*. Phytochemistry 66:1254–1263

42. Hemmati S, Schmidt TJ, Fuss E (2007) (+)-Pinoresinol/(-)-lariciresinol reductase from *Linum perenne* himmelszelt involved in the biosynthesis of justicidin B. FEBS Lett 581:603–610

43. Umezawa T, Davin LB, Yamamoto E et al (1990) Lignan biosynthesis in *Forsythia* species. J Chem Soc Chem Commun:1405–1408

44. Umezawa T, Davin LB, Lewis NG (1991) Formation of lignans (-)-secoisolariciresinol and (-)-matairesinol with *Forsythia intermedia* cell-free extracts. J Biol Chem 266:10210–10217

45. Okunishi T, Sakakibara N, Suzuki S et al (2004) Stereochemistry of matairesinol formation by *Daphne* secoisolariciresinol dehydrogenase. J Wood Sci 50:77–81

46. Xia Z-Q, Costa MA, Pélissier HC et al (2001) Secoisolariciresinol dehydrogenase purification, cloning, and functional expression. J Biol Chem 276:12614–12623

47. Youn B, Moinuddin SGA, Davin LB et al (2005) Crystal structures of apo-form and binary/ternary complexes of *Podophyllum* secoisolariciresinol dehydrogenase, an enzyme involved in formation of health-protecting and plant defense lignans. J Biol Chem 280:12917–12926

48. Moinuddin SGA, Youn B, Bedgar DL et al (2006) Secoisolariciresinol dehydrogenase: mode of catalysis and stereospecificity of hydride transfer in *Podophyllum peltatum*. Org Biol Chem 4:808–816

49. Sakakibara N, Suzuki S, Umezawa T et al (2003) Biosynthesis of yatein in *Anthriscus sylvestris*. Org Biol Chem 1:2474–2485

50. Katayama T, Ogaki A (2001) Biosynthesis of (+)-syringaresinol in *Liriodendron tulipifera* I: feeding experiments with L-[U-^{14}C]phenylalanine and [8-^{14}C]sinapyl alcohol. J Wood Sci 47:41–47

51. Rahman MA, Katayama T, Suzuki T et al (2007) Stereochemistry and biosynthesis of (+)-lyoniresinol, a syringyl tetrahydronaphthalene lignan in *Lyonia ovalifolia* var. *elliptica* II: feeding experiments with ^{14}C labeled precursors. J Wood Sci 53:114–120

52. Liu J, Stipanovic RD, Bell AA et al (2008) Stereoselective coupling of hemigossypol to form (+)-gossypol in moco cotton is mediated by a dirigent protein. Phytochemistry 69:3038–3042

53. Umezawa T, Okunishi T, Shimada M (1997) Stereochemical diversity in lignan biosynthesis. Wood Res 84:62–75

54. Umezawa T (2001) Biosynthesis of lignans and related phenylpropanoid compounds. Regul Plant Growth Dev 36:57–67

55. Nakatsubo T, Mizutani M, Suzuki S et al (2008) Characterization of *Arabidopsis thaliana* pinoresinol reductase, a new type of enzyme involved in lignan biosynthesis. J Biol Chem 283:15550–15557

56. Gottlieb OR (1972) Chemosystematics of the lauraceae. Phytochemistry 11:1537–1570

57. Moinuddin SGA, Hishiyama S, Cho M-H et al (2003) Synthesis and chiral HPLC analysis of the dibenzyltetrahydrofuran lignans, larreatricins, 8′-*epi*-larreatricins, 3,3′-didemethoxyverrucosins and *meso*-3,3′-didemethoxynectandrin B in the creosote bush (*Larrea tridentata*): evidence for regiospecific control of coupling. Org Biomol Chem 1:2307–2313

58. Lopes NP, Yoshida M, Kato MJ (2004) Biosynthesis of tetrahydrofuran lignans in *Virola surinamensis*. Brazil J Pharm Sci 40:53–57

59. Messiano GB, da Silva T, Nascimento IR et al (2009) Biosynthesis of antimalarial lignans from *Holostylis reniformis*. Phytochemistry 70:590–596

60. Vassão DG, Gang DR, Koeduka T et al (2006) Chavicol formation in sweet basil (*Ocimum basilicum*): cleavage of an esterified C9 hydroxyl group with NAD(P)H-dependent reduction. Org Biomol Chem 4:2733–2744

61. Koeduka T, Fridman E, Gang DR et al (2006) Eugenol and isoeugenol, characteristic aromatic constituents of spices are biosynthesized via reduction of a coniferyl alcohol ester. Proc Natl Acad Sci USA 103:10128–10133
62. Koeduka T, Baiga TJ, Noel JP et al (2009) Biosynthesis of *t*-Anethole in anise: characterization of *t*-Anol/Isoeugenol synthase and an *O*-Methyltransferase specific for a C7-C8 propenyl side chain. Plant Physiol 149:384–394
63. Dexter R, Qualley A, Kish CM et al (2007) Characterization of a petunia acetyltransferase involved in the biosynthesis of the floral volatile isoeugenol. Plant J 49:265–275
64. Umezawa T (2003) Phylogenetic distribution of lignan producing plants. Wood Res 90:27–110
65. Tazaki H, Hayashida T, Ishikawa F et al (1999) Lignan biosynthesis of liverworts *Jamesoniella autumnalis* and *Lophocolea heterophylla*. Tetrahedron Lett 40:101–104
66. Orr JD, Lynn DG (1992) Biosynthesis of dehydrodiconiferyl alcohol glucosides: implications for the control of tobacco cell growth. Plant Physiol 98:343–352
67. Sartorelli P, Benevides PJC, Ellensohn RM et al (2001) Enantioselective conversion of *p*-hydroxypropenylbenzene to (+)-conocarpan in *Piper regnellii*. Plant Sci 161:1083–1088
68. Katayama T, Kado Y (1998) Formation of optically active neolignans from achiral coniferyl alcohol by cell-free extracts of *Eucommia ulmoides*. J Wood Sci 44:244–246
69. Lourith N, Katayama T, Ishikawa K et al (2005) Biosynthesis of a syringyl 8-*O*-4′ neolignan in *Eucommia ulmoides*: formation of syringylglycerol-8-*O*-4′-(sinapyl alcohol) ether from sinapyl alcohol. J Wood Sci 51:379–386

Index

D.R. Gang (ed.), *The Biological Activity of Phytochemicals*, Recent Advances
in Phytochemistry 41, DOI 10.1007/978-1-4419-7299-6,
© Springer Science+Business Media, LLC 2011

Autosampler, 99
Avenacin
    fluorescence, 15
    model, 15
Avenacin A-1, 14–16, 20
Avena strigosa (wild oat), 15
AVIs, *see* Anthocyanic vacuolar inclusions
    (AVI)

B
Baker, B. J., 1–11
Benzyl glucosinolate, 147
Benzyl mercaptan, 40
Bernards, M. A., 13–28
Bidesmosidic saponins, 14–19
Bifidobacterium breve, 25
Biomek 3000, 11
Biosynthesis of carotenoids in peppers
    desaturation reactions, 113
    mevalonic acid pathway, 112
    plastidal pathway, 112
    rate-limiting enzyme
        phytoene synthase, 114
Biosynthesis of flavonoids in citrus
    anthocyanidin synthase
        analysis, 79
        identification, 79
        leucoanthocyanidins to anthocyanidins,
            conversion, 79
    chalcone isomerase
        biochemically characterized, 75
        CHI activity identification, 75
        isolation, 75
        isoliquiritigenin activity, 75
        purification, 74
        stereospecific cyclization, 74
    chalcone synthase
        characteristics, 74
        CHS enzyme activity levels, 74
        expression of CitCHS2, 74
        nucleotide sequence, 73
        reaction mechanism, 73
    cinnamate 4-hydroxylase
        purification, 72
    4-coumarate: CoA ligase
        biochemical characterization, 73
        multiple isozymes, 73
    dihydroflavonol 4-reductase
        blood orange cultivars, 78
        expression of, 78
        Southern blot analysis, 78
        stereospecific conversion, 78
    flavanone 3β-hydroxylase

characterization, 76
    cofactors identified in flavonoid
        pathway, 76
    conversion of naringenin to
        dihydrokaempferol, 76
    isolation, 76
    flavanone 3′–hydroxylase
        activities, 79
        cofactors, 79
        identified by, 79
    flavanone 3′,5′–Hydroxylase
        extracts of *Verbena hybrida* flowers, 80
        identification of, 80
    flavone synthase
        activity identification, 76
        flavone and flavanone glycosides,
            accumulation, 77
    flavonol synthase
        biochemical and structure/function
            analysis, 77
        catalyzation, 77
        identification, 77
        Southern blot analysis, 77–78
    phenylalanine ammonia lyase
        investigation results, 72
        purification, 69
        Southern blot analysis confirmation, 72
        TAL activity mechanism, 69–70
Biosynthesis
    lignan
        neolignans, biosynthesis, 191–193
        other types, 190–191
        with 9(9′)–oxygen, 184–190
    norlignan
        A. officinalis, 182
        enantiomeric composition, 183
        hinokiresinol, 180
        hypothetical biosynthetic pathway, 181
        mechanism, 182
        product biosynthesis, 182
        recombinant proteins, 182
    norlignans/lignans/neolignans, basic
        carbon skeletons, 180
    phenylpropanoid derivatives,
        functions, 179
BioTek HT plate reader, 10
Biswasalexins A1, 132–133
Bosland, P. W., 109–121
Botrytis cinerea, 18, 20, 25
Bovine serum albumin (BSA), 10, 99
Brassinin oxidase/dehydrogenase,
    135–136
Brian Miki, B., 47–55